I0023266

*Metamorphosis:*
**The Case for Intelligent Design**
**In a ~~Nutshell~~ Chrysalis**
**A Companion Book to the Film**

Edited by David Klinghoffer

Published by Discovery Institute Press, 208 Columbia Street, Seattle, WA 98104, United States of America.

We are the caterpillars of angels.
—Vladimir Nabokov

# Table of Contents

## A Special Message from Dean Koontz:

**Metamorphosis** is dazzling, insightful, and thought-provoking, with the power to open closed minds. For those of us who love science but decry scientism, this film affirms what we see everywhere from the latest discoveries in molecular biology to the long-understood twenty universal constants that, in their exquisite balance, make life possible: intention, meaning, and an intricacy that confounds all theories portraying nature as a consequence of dumb forces.

Dean Koontz

*http://www.deankoontz.com*

*Hailed by* Rolling Stone *as "America's most popular suspense novelist,"
Dean Koontz has been compared to Flannery O'Connor, Walker Percy,
and Charles Dickens. His books have been published in 38 languages
and have sold more than 400 million copies.*

**Section I: Introducing *Metamorphosis***

# Chapter 1
## About This Book:
## A Still Small Voice

David Klinghoffer

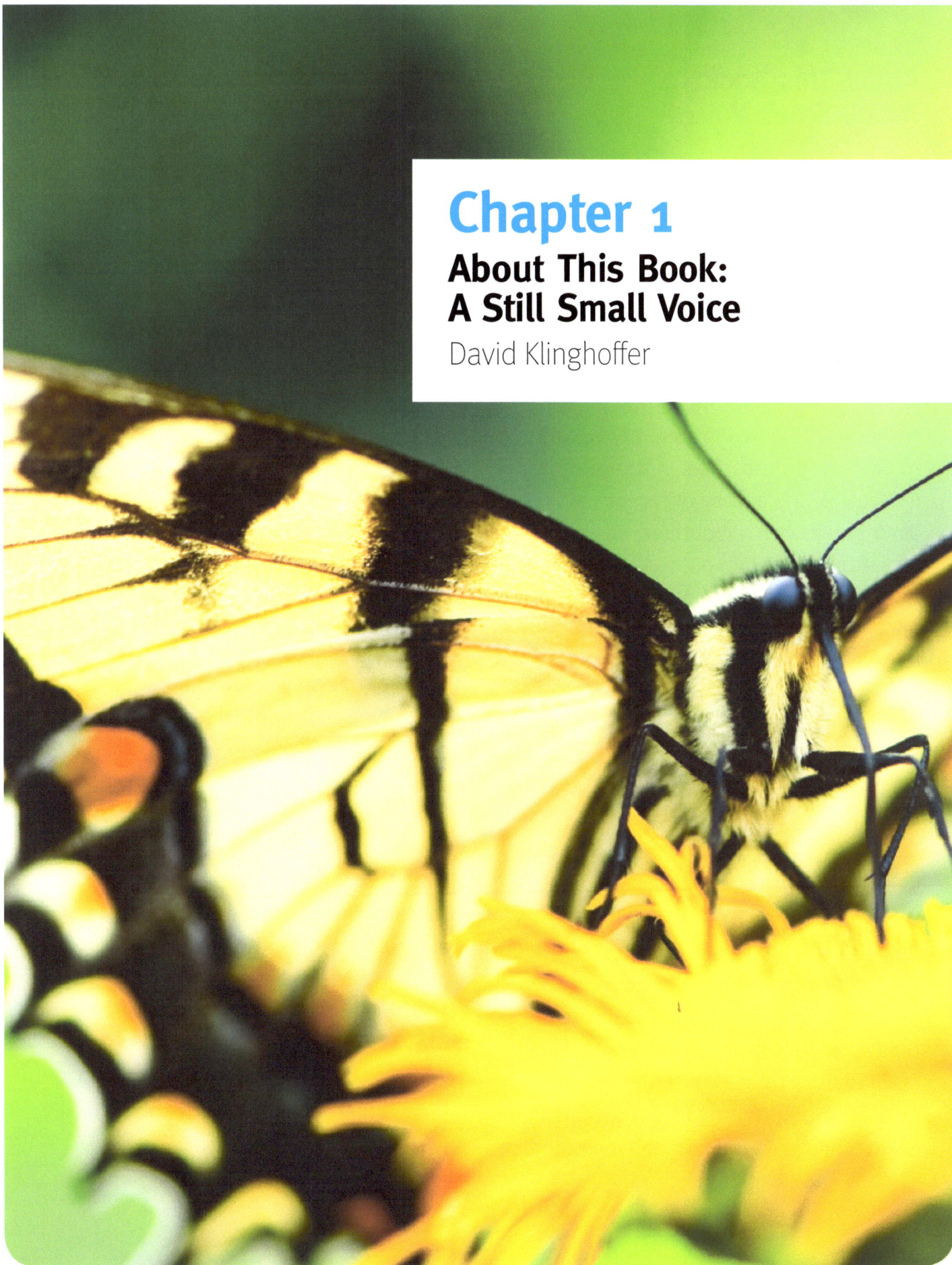

# Chapter 1
## About This Book:
## A Still Small Voice

David Klinghoffer

The purpose of this book is to serve as a companion to the Illustra Media documentary *Metamorphosis* for readers who want to know more about butterflies and the challenges they pose to Darwin's theory. In editing it and writing some chapters, I thought with regret about how infrequently we get to see living butterflies in the wild here in Seattle. It's a very different environment from my Southern California hometown which even had a unique native butterfly named after it, the now endangered Palos Verdes Blue *(Glaucopsyche lygdamus palosverdesensis)*. Trying to remember the last time I saw a butterfly in a natural setting, I drew a blank.

The butterfly season in Washington State is brief, unlike California where it is year-round. Butterflies favor sunny, warm weather and generally only come out when it's over 60 degrees. They don't care for the surroundings of a city and even a suburb lacks the comforts they find most agreeable. In *Butterflies of the Pacific Northwest*, William Neill advises to seek them in open pine forests, mountain meadows and alpine ridges, high desert river canyons, and

"Neglected, weedy roadsides wherever you find them."[1]

None of that sounds much like the neat if wooded suburb where our family lives. So on a sunny Saturday in June, I counted it as an unexpected blessing that a neighbor of ours had let the border of property fronting his home turn to shrubby flowering weeds. On the way home from synagogue, our four-year-old twin Saul and I were walking a steeply descending street, running down toward the edge of Lake Washington. It was lovely spring scene and there, topping it off, was a most beautiful butterfly.

Yellow with black stripes extending back from the edge of its forewing, a good three and a half inches across, very large for a butterfly, it had swooped in for a sip of nectar, sharing the tiny flowers of the neglected shrub with a group of fuzzy warm orange bumblebees who were seeking their own livelihood. Later I identified it as a Western Tiger Swallowtail *(Papilio rutulus)*, probably a female judging from the prominent iridescent blue spots along the submarginal area of the hind wings.

Catching sight of the swallowtail, I drew Saul up for a closer look. I explained to him that the appendage on the front called the proboscis, a bit like

---

1 William Neill, *Butterflies of the Pacific Northwest* (Missoula, Montana: Mountain Press, 2007), p. 26.

the straw that comes with those little juice boxes kids like to get in their lunch, is how the butterfly obtains her favorite beverage. Nectar is something like a sugary energy drink that kids would also enjoy.

My son looked on with interest. "First he has to eat food," Saulie commented. "And then when he's done eating, he says to the bees 'Can you play with us?'"

Figure 1: The Western Tiger Swallowtail.

**"** *My son looked on with interest. "First he has to eat food," Saulie commented. "And then when he's done eating, he says to the bees 'Can you play with us?'"* **"**

Later I told our older kids that the Western Tiger Swallowtail in its early larval stage disguises itself by mimicking a bird dropping. They found this hilarious and the occasion for an outburst of potty talk. I didn't mention that many butterflies get nutrition other than nectar by alighting on dog or coyote scat from which they draw amino acids, fats and minerals. To each his own!

The butterfly let us look at her for about 15 seconds and then flew away to the next shrub a few feet down the street. Saulie and I ran after her. This broke what I later learned is a cardinal rule of stalking butterflies: Don't run after them. You are also supposed to walk softly and thus avoid setting off vibrations she can feel through her legs. So as not to scare her away, also avoid casting a shadow over the insect, and try to somehow blend the outline of your body with that of the shrub. We broke all these rules too.

The swallowtail fluttered delicately, awkwardly, down the street to another shrub, and we ran after her again. Then she was off to the next shrub, and then away again, at which point we decided to halt our pursuit. At any time while she was feeding, I could easily have reached out a hand and caught her. Later I wondered, as have others, how such a vulnerable, defenseless, conspicuous and leisurely creature emerged, if the conventional evolutionary scenario is believed, as the fittest to survive over competitors.

We had the opportunity to closely observe this grand insect for about a minute all told. It was a minute to treasure in memory. Paul A. Opler in *A Field Guide to Western Butterflies* offers that when asked the ever popular question of what distinguishes butterflies from moths, both groups in the order Lepidoptera, he explains that "butterflies are really just part of the vast evolutionary variation in the order. Another way to put it is to say that butterflies are just 'fancy moths.' Butterflies have become popular partly because they are conspicuous and be-

cause there are neither too few nor too many species to pique our interest."[2]

This understates, by a large measure, the spell of enchantment cast by butterflies. Even to speak of their beauty would miss the mark. Despite all the undoubted beauty of that spring scene, with the trees, the lake, and the green highlands rising in the distance under a warm sun, you might still, if you really insisted, take it all as the fortunate production of an unguided process of cosmic churning, the same that produces stars and planets, oceans and deserts, and ultimately Darwin's tree of life. But now add that swallowtail to the scene. It is a fluttering signifier of art and artifice if ever there was one. Dismissing nature as the product of blind, seething forces has just got a lot harder to do.

As at least my own tradition would have it, the supernal wisdom that pervades, underlies, and maintains existence is necessarily obscured and concealed from us. We are like the visitor to a museum, gazing at a painting on a canvas and getting lost in its fictional reality, forgetting that it is the projection of an artist's mind and creativity. The artist and the viewer collaborate in this agreeable illusion. At certain moments, however, it is as if we are that same museumgoer when he notices a

2 Paul A. Opler, *A Field Guide to Western Butterflies*, illustrated by Amy Bartlett Wright (Boston: Houghton Mifflin, 1999), p. 4.

signature in the corner of the canvas, that of the artist. This breaks the illusion that what we are seeing on the museum wall could be an actual scene of life in the world, something glimpsed by looking in or out of a window. No, the painting is an artifice, the production of a designer.

For the museumgoer, the realization may come as a disappointment. It's almost, but not quite, like the let down you feel when the lights come on in a movie theater after the show and a young person from the theater staff starts going around with a garbage bag, picking up litter.

Seeing a butterfly, of course, is the opposite of disappointing. The indication that the canvas of nature bears such a mark of authorship, one among many other signs, is one of those experiences in life that give you hope, in a culture blighted by cynicism, that the enchantment we sometimes feel is no illusion after all. On the contrary, it points to the ultimate reality, lying only just behind the reality we observe. A butterfly dancing in the sunlight is a finger tapping you gently on the shoulder, a still small voice from somewhere behind saying, "Don't be fooled."

A butterfly is not unique in this. You could probably think of many moments in your own life where catching sight of something unexpected caused the

scales to briefly fall away. There are as many as there are species of actual butterflies, "neither too few nor too many to pique our interest," as Paul A. Opler might say. Whittaker Chambers described in his 1952 memoir, *Witness*, the moment he awoke from his earlier Communism: It was upon looking closely one day at his young daughter's ear. He was feeding her oatmeal and even as the food got on her face and on the table top, he noted the exquisite beauty of the tiny ear and the evidence of "immense design" it gave.[3] The experience shook him. He could never again subscribe to the secular, materialist dream.

It could be something as small as an ear or as great as an ocean. Before my father passed away this year, he was ill for months in a Los Angeles hospital. There was an occasion when, on a visit to see him, I concluded a hard day by driving out to the beach by the Santa Monica Pier. With sometimes high levels of chemical and bacterial pollution, Santa Monica is no pristine beach paradise. Yet somehow, standing with bare feet in the questionable water, the vastness of the Pacific abruptly calmed and cheered me, dispelling a darkness. Contact with death and dying makes us vulnerable to the feeling that we are helpless material beings in the grasp

---

[3] Whittaker Chambers, *Witness* (Washington, D.C.: Regnery Publishing, 1987), p. 16.

of a mindless material universal. It is a condition expressed in the Hebrew Bible as *tumah* or ritual contamination for which the prescribed remedy is immersion in "living water." Touching the ocean, even at a somewhat polluted urban beach, counteracts the contaminating illusion. Like a butterfly, it leaves us surprised, grateful and wanting more. It whispers, "Don't be fooled."

This is an intuition and an intuition can be mistaken. Some observations need to be proved but some don't. In his wonderful little book *Real Presences*, the literary critic and philosopher George Steiner teaches that if materialism were to really win the day and conquer our culture, the expression and recognition of beauty would be crippled. He calls this a conjecture and admits it can't be proven. Yet Steiner also cites Aristotle's *Metaphysics* to the effect that knowing when an idea needs to be proven at all is a matter of *apaideusis*. The Greek word can "be translated as meaning a want of schooling, a fundamental lesion in education. I would render the term as connoting an indecency of spirit and of understanding."[4]

To feel unmoved on seeing a butterfly, or even to feel moved yet to ask for harder proof that the creature points to the presence of an invisible reality behind nature, may well be indecent. If so, then you can call me indecent. *Metamorphosis* is a fantastically beautiful and informative documentary, but it left me hungry for a more detailed and conclusive treatment of the contradictions that butterflies have long been recognized as posing to Darwinian materialist philosophy. We have gathered the essays in this book because doubtless many other viewers will want, if not proof, then at least an elaboration of the ideas to which the film briefly alludes.

---

[4] George Steiner, *Real Presences* (Chicago: University of Chicago Press, 1991), p. 231.

# Chapter 2

## Reviewing *Metamorphosis*: The Case for Intelligent Design in a ~~Nutshell~~ Chrysalis

David Klinghoffer

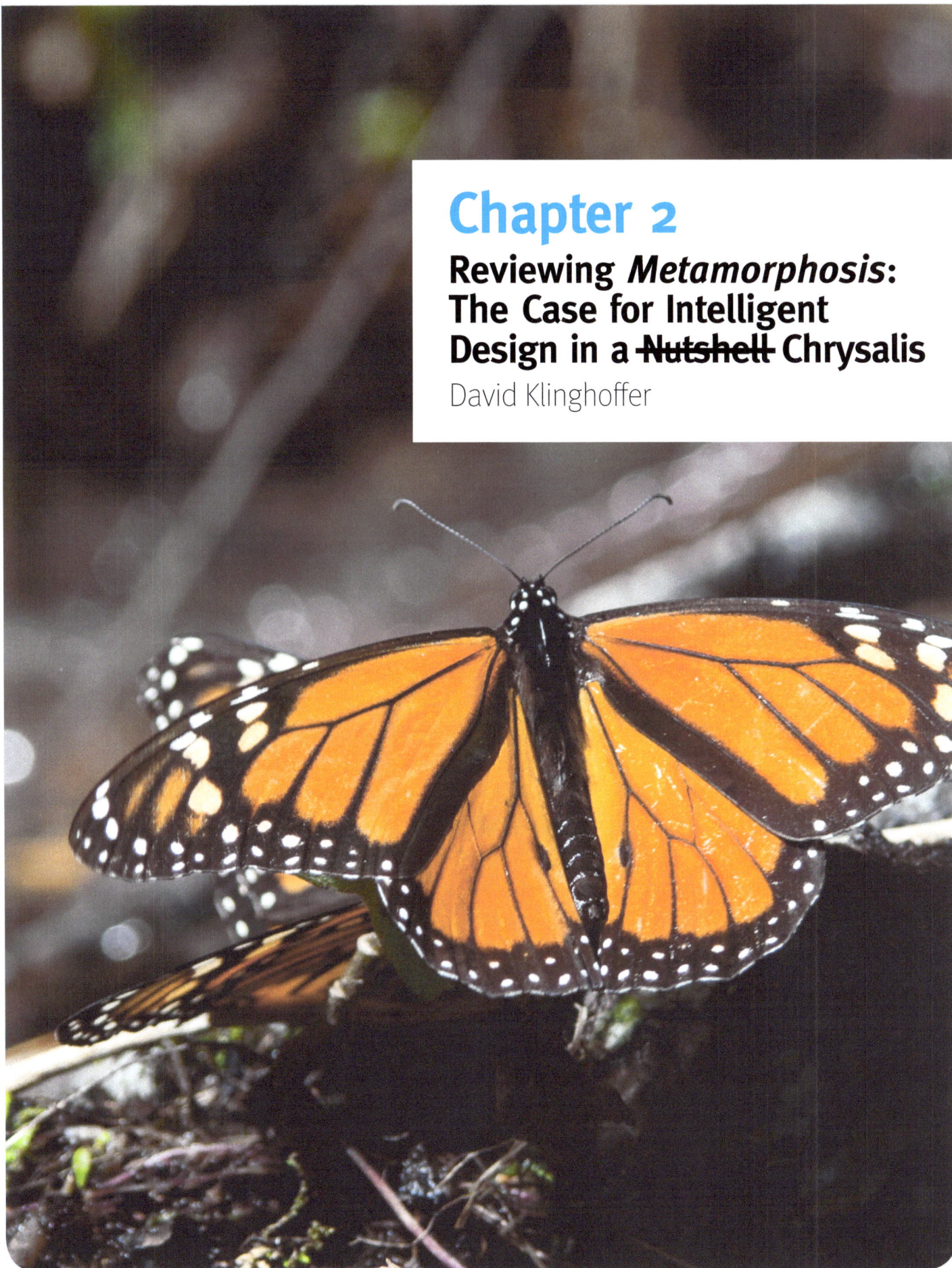

## Chapter 2
# Reviewing *Metamorphosis:* The Case for Intelligent Design in a ~~Nutshell~~ Chrysalis

David Klinghoffer

The other night, I watched the latest production from Illustra Media, *Metamorphosis*, with our oldest kid, nine-year-old Ezra. Given that he pretty strictly requires that video entertainment involve robots flying around blowing things up, I expected him to scoff at a movie about caterpillars that crawl around, turn into butterflies then proceed to fly to Mexico. Conspicuously, on its remarkable unguided cross-continental journey, the luminous orange-and-black Monarch butterfly fails to blow up anything at all.

Yet Ezra sat entranced throughout, as I did, which leads me to think *Metamorphosis* is going to be a big, cross-generational hit.

*Metamorphosis* follows on the heels of past Illustra offerings, including *Privileged Planet, Unlocking the Mystery of Life*, and *Darwin's Dilemma*. It's probably true that with these films taken altogether, Illustra producer and documentarian Lad Allen has made the most easily accessible, visually stunning case for intelligent design available.

If you have one shot at opening the mind of an uninformed and dismissive friend or family member, the kind who feels threatened by challenges to Darwinism, then presenting him with a copy of a 600-page volume like Stephen Meyer's *Signature in the Cell*, or even a slimmer alternative like Michael Behe's *Darwin's Black Box*, would probably be less effective than choosing one of Mr. Allen's DVDs. Among those, *Metamorphosis* might well make the best initial selection, since the argument for intelligent design doesn't come in till the third and final act. When it comes, it's a soft sell, preceded by a gorgeous, non-threatening nature film that only hints at what's ahead in Act III. In Act I, the focus is on the mind-blowing magical routine by which the caterpillar enters into the chrysalis, dissolves into a buttery blob and swiftly reconstitutes itself into a completely different insect, a butterfly.

A cute graphic sequence shows, by way of analogy, a Ford Model T driving along a desert road. It screeches to a stop and unfolds a garage around itself. Inside, the car quickly falls to pieces, divesting itself of constituent parts that spontaneously recycle themselves into an utterly new and far more splendid vehicle. A sleek modern helicopter emerges from the garage door and thumps off into the sky.

Figure 2: The Monarch making its journey of migration.

*"Ancient philosophers and mystics spoke of an "animal soul," different from the soul that makes human beings unique..."*

In Act II, we follow a particular butterfly, the Monarch, on its journey to a volcanic mountain lodging site in Mexico for the winter, accomplished each year despite the fact that no single, living Monarch was among the cohort that made the trip the year before. Only distant relations—grandparents, great-grandparents—did so. Given the brief life cycle of the insect, those elders are all dead. The Monarch follows the lead of an ingenious internal mapping and guidance system dependent on making calculations of the angle of the rising sun and on magnetic tugs from ferrous metal in the target mountain range.

Experts explain and comment, including Center for Science & Culture fellow and philosopher of biology Paul Nelson, Biologic Institute developmental biologist Ann Gauger, and University of Florida zoologist Thomas Emmel. The film argues that neither metamorphosis nor migration is the kind of feature with which blindly groping Darwinian natural selection could ever equip a creature. How could an unguided step-by-step process build metamorphosis, inherently an all-or-nothing proposition? As Dr. Gauger points out, once the caterpillar has entered the chrysalis, there's no going back. It must emerge either as a fully formed butterfly or the soupy remains of a dead caterpillar.

If I had a criticism of the film, it would be that too little time is devoted to the evolution debate. You come away wondering how Darwinists would respond, and how ID-friendly experts would reply in turn. Hence part of our reason for publishing this book.

Well, Lad Allen's film won't be the last word on the subject, just as it is far from the first. Contemplating butterflies was among the considerations that drove evolutionary theory's co-discoverer, Alfred Russel Wallace, to doubt the sufficiency of natural selection to account for the most wondrous aspects of animal life. Like lepidopterist and novelist Vladimir Nabokov a half-century later, Wallace noted the astonishing, gratuitous artistry with which butterflies adorn their wings.

In *The World of Life*, Wallace wrote of how he could satisfyingly account for this only as a feature intended by design "to lead us to recognize some guiding power, some supreme mind, directing and organizing the blind forces of nature in the production of this marvelous development of life and loveli-

ness." Butterflies may not literally blow up bad guys like the robots in my son's favorite movies, but they strike another blow for Wallaceism.

More subtly, the transformation of the caterpillar hints at a deeper truth about life, that it is not bestowed on machines or other mechanical devices, as per the mechanistic myth. Ancient philosophers and mystics spoke of an "animal soul," different from the soul that makes human beings unique, although people possess both an animal and a divine soul, along with our physical bodies. The animal soul, in this view, is a vital force received by inheritance at conception and, among other functions, participating in the direction of how the body gets knitted together.

Speaking of it as a soul implies purpose, intention, intelligence. That sure does look like what's at work in those mere couple of weeks spent in the chrysalis. Darwinism, of course, has a hard enough time explaining the construction of a living machine. This is something much greater, posing a far harder challenge to materialist evolutionism.

# Chapter 3
## An Interview with Lad Allen, Producer and Director of *Metamorphosis*

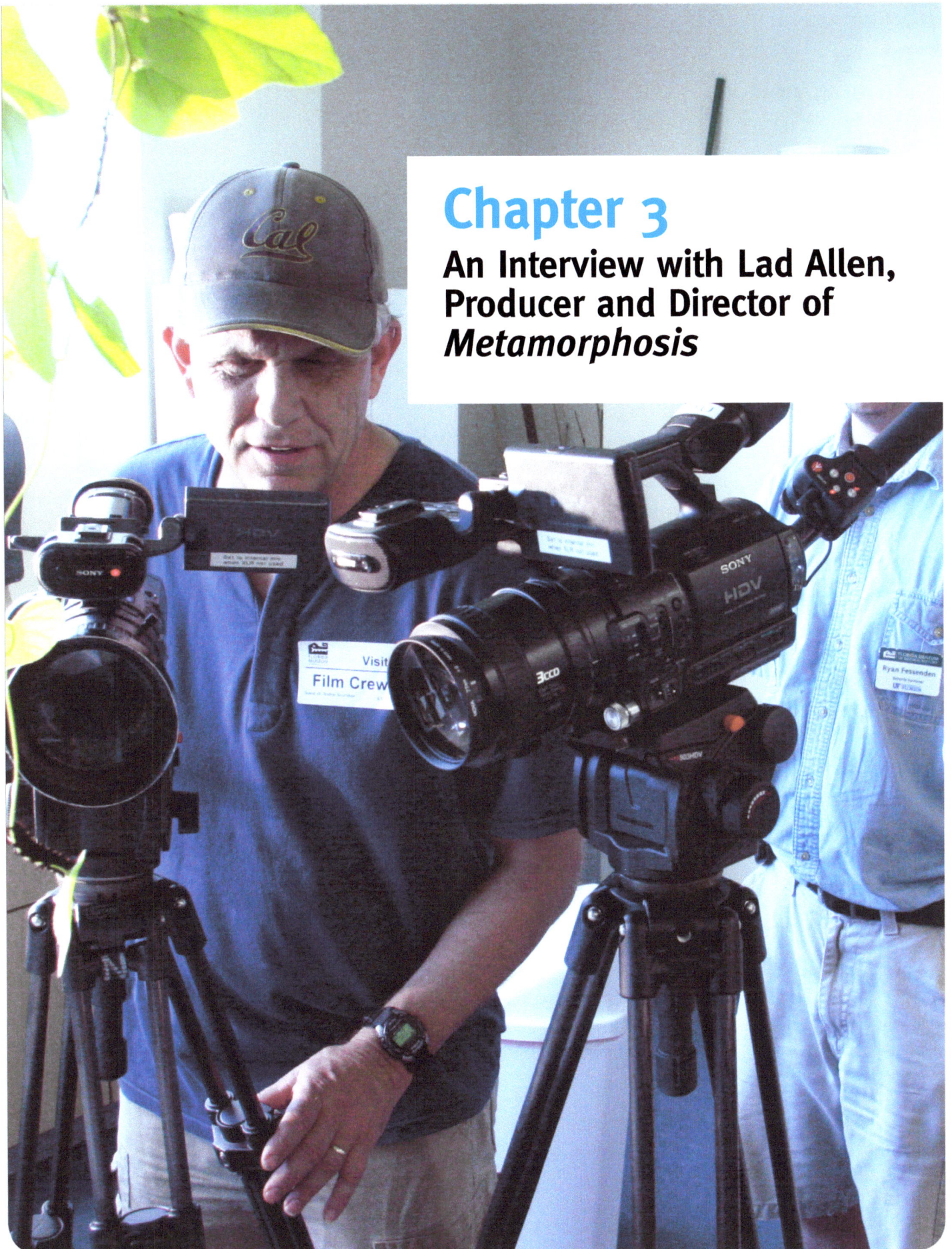

# Chapter 3
# An Interview with Lad Allen, Producer and Director of *Metamorphosis*

Lad Allen has been making films exploring the wonders of nature for three decades. Since 2002, he and editor Jerry Harned of Illustra Media have been responsible for creating a series of outstanding science documentaries about the evidence for intelligent design in the universe: *Unlocking the Mystery of Life*, *The Privileged Planet*, *Darwin's Dilemma*, and now *Metamorphosis*. The first two films have been shown on dozens of PBS stations around the United States, and *The Privileged Planet* was screened at the Smithsonian's National Museum of Natural History in Washington, D.C. Translated into more than two dozen languages, Illustra Media's documentaries are known around the world for their high production values, their engaging content, and their cutting-edge science. Here Lad Allen talks about the story behind the creation of his latest film, *Metamorphosis*.

## What is it about butterflies that you find so fascinating?

**ALLEN:** I find butterflies fascinating for several reasons. They are spectacularly beautiful—among the most beautiful creatures on the planet. The variety of colors, patterns, and wing shapes are extraordinary. They are also marvels of engineering. Their wings are covered with thousands of microscopic solar panels that warm the cold-blooded insects for flight. Their senses of smell (with their antennae) and taste (with their feet) are highly developed. Their compound eyes create a field of vision more than 180 degrees wide. Their wings adjust, in flight, to take advantage of even the slightest changes in wind currents. Their life cycle (egg to larva to chrysalis to adult) is one of the most mysterious and miraculous transformations in nature.

## What is the back-story of the film? How did you come to do this project?

**ALLEN:** In 1988-89, we produced two films about butterflies for Callaway Gardens and the Cecil B. Day Butterfly Center in Atlanta. At the time, I was working for Moody Institute of Science. These projects really hooked me on butterflies. We photographed hours of wonderful footage in Mexico, Ecuador, and at the Day Butterfly Center. In 1997, Moody decided to curtail its film production work. All of the butterfly footage was shipped to Chicago, where it was stored in a basement for more than a decade. In 2009, I received a call asking if we would be interested in the footage. Jerry Harned and I flew back to Chicago, found the butterfly material, and had it shipped back to California. It became the foundation of *Metamorphosis*. In January 2010, we decided to proceed with production of a documentary that would explore evidence for intelligent design based upon the life cycle of butterflies and the epic migration of the Monarch butterfly. We felt the story and subject matter would have universal appeal, and that the evidence for design was compelling.

*"Metamorphosis proved to be one of the most challenging and enjoyable films we have ever made."*

Figure 3: Jerry Harned on location.

### What was it like making this film? What are some of the locations you used?

**ALLEN:** *Metamorphosis* proved to be one of the most challenging and enjoyable films we have ever made. We traveled to the Transvolcanic Mountain Range in Mexico (elevations over 10,000 feet) to film the winter sanctuary for most of the North American Monarch butterfly population (more than a billion butterflies that migrate to Mexico from as far north as Canada). It was thrilling to stand in a forest with hundreds of thousands of Monarchs flying around you on a sunny morning. You could hear their wings flap.

We also filmed extensively at the McGuire Center for Lepidoptera at the University of Florida, and at Butterfly World in Ft. Lauderdale. The McGuire Center is one of the finest butterfly research centers in the world. Butterfly World is the largest display of free-flying butterflies (about 10,000 in all) in the Western hemisphere. There, we photographed every stage of the metamorphosis process, and slow-motion studies of butterflies in flight.

**Butterflies often inspire a sense of wonder in children and adults alike. Did you experience that when working on the film?**

**ALLEN:** This entire project was enveloped in wonder. A butterfly's life cycle is still one of the great mysteries of the natural world. An earth-bound caterpillar encases itself in a casing called a chrysalis. There, its organs are dissolved into a chemical soup. They are then rearranged to help build wings, compound eyes, reproductive systems, and a host of other organs that did not exist in the caterpillar. It's an incredible process that screams out purpose, foresight, engineering, and design. Every scientist and scholar we interviewed was in awe of the process.

The same is true for the migration of the Monarch butterflies. Monarchs that emerge in the spring or early summer live for about two to four weeks. But the generation that emerges in late August is genetically equipped to live up to nine months. It's called the "Methuselah Generation." This enables these tropical butterflies (that would die if exposed to the freezing winter temperatures of the Midwest and Canada) to migrate as far as 3,000 miles, to a small area of forest in the Transvolcanic Mountains of central Mexico. There, the conditions are right to ensure the survival of the Monarchs until spring. In March, the Monarchs become sexually active for the first time. They mate and then begin their return migration north. When they reach southern Texas, the females lay their eggs (only on milkweed plants—the only food source their caterpillars will eat) and soon die. Throughout the summer months, new generations of Monarchs emerge and move north—living, again, between two and four weeks. Then, in early September, a new Methuselah Generation—three or four generations removed from the Monarchs that migrated the previous year—travel from as far north as Canada to the same trees that provided sanctuary for their grandparents and great grandparents, the year before. The navigational systems that enable these insects (that each weigh less than a quarter of an ounce) to navigate so precisely to a forest in Mexico they have never been before are incredible.

### How do butterflies help make a case for intelligent design?

**ALLEN:** I think the beauty evident in butterflies is evidence for design. The patterns, colors, and shapes of their wings are far beyond what is required for camouflage or attracting a mate. I think it's often a case of gratuitous, over-the-top beauty that may exist—in some measure—for human beings to enjoy. Natural selection, by definition, doesn't lead to physical characteristics that exist for purposes other than survival. It doesn't make things for the purpose of "just being beautiful." But intelligence does.

You also see strong evidence for design in the metamorphosis process itself. When the caterpillar enters the chrysalis stage it kills itself, as its internal organs dissolve. Before it initiates that stage of its life cycle, it has to have a plan for getting out the other side as a fully functioning adult. If the transformation is incomplete, the caterpillar is dead.

This fact makes a Darwinian explanation for metamorphosis extremely difficult to formulate. Before there was a butterfly, how did the first caterpillars rebuild their bodies—into winged insects that didn't previously exist— through a series of small gradual evolutionary steps...*that included cell death and suicide?* The only way this would be possible is if metamorphosis was orchestrated by an agent capable of foresight and purpose. An agent that could look into the future and visualize what it was going to become. An undirected natural process couldn't do that. But an intelligent designer could.

# Section II: Butterflies and the Case for Intelligent Design

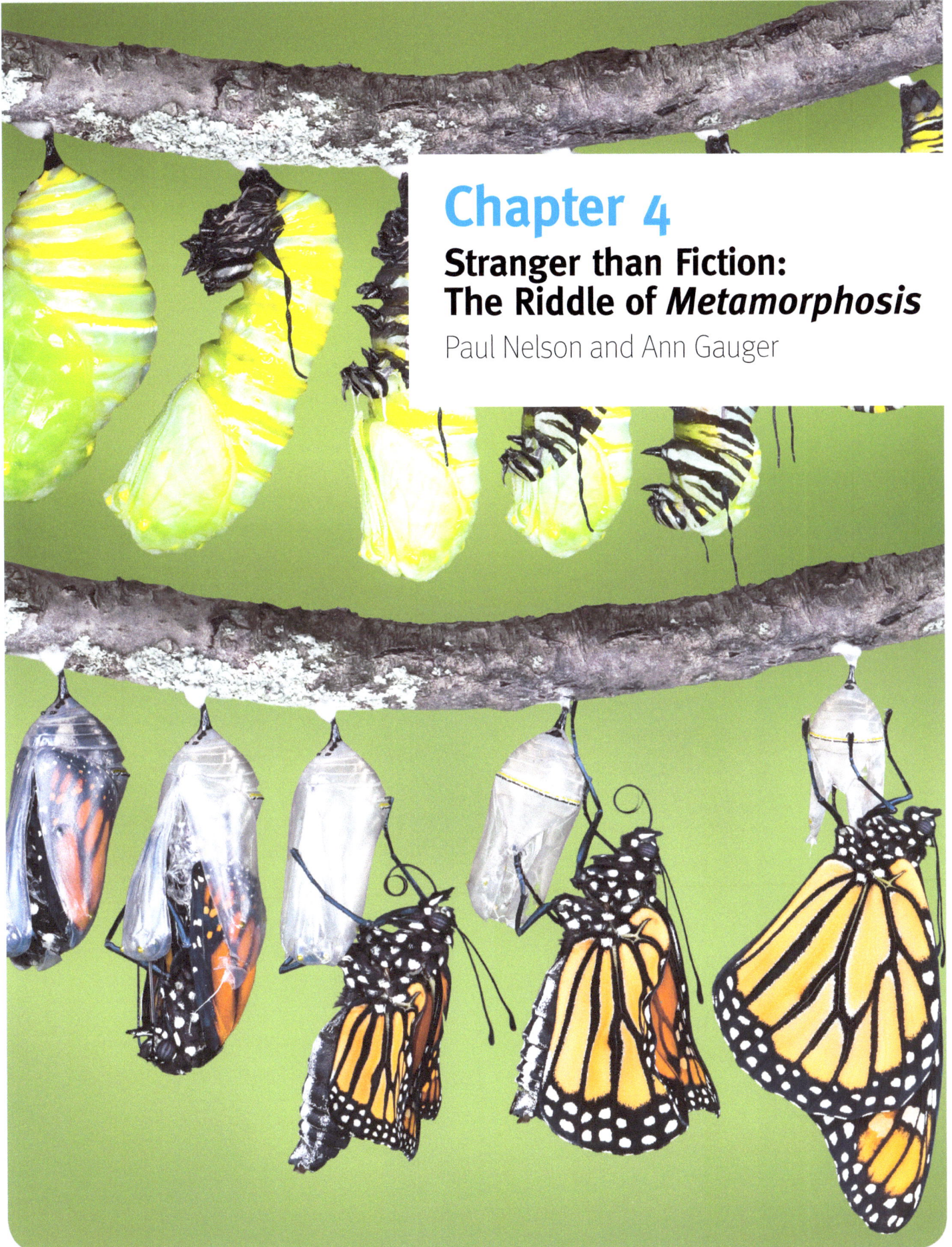

# Chapter 4
## Stranger than Fiction: The Riddle of *Metamorphosis*
Paul Nelson and Ann Gauger

# Chapter 4
## Stranger than Fiction:
## The Riddle of *Metamorphosis*
Paul Nelson and Ann Gauger

Imagine tucking yourself into bed one night and waking up as a completely different creature. Your tissues have dissolved, and you have grown a new skeleton, a new straw-like mouth, eyes capable of seeing colors into the ultraviolet and near infrared, a new and powerful sense of smell, and an iridescent set of wings. You can eat only nectar, and you are overwhelmed by the urge to find another creature like you and mate with him or her. Sound like fiction?

Yet this is the life story of the caterpillar. It hatches out of the egg as a worm-like creature whose sole purpose is to eat as much as possible and to grow as rapidly as it can. When it has grown large enough, it tucks itself into bed (we know it as the chrysalis or pupa), and over the course of several days, while it sleeps, so to speak, its body is built anew. Caterpillar (larval) tissues are dissolved or remodeled, and new wings, legs, eyes, antennae, nerve connections, muscles, epidermis, and reproductive organs develop. Even the brain itself undergoes a substantial transformation. The adult butterfly finally emerges as a beautiful, free-flying animal, completely unlike what came before.

Thus the butterfly's life history involves the development of not one, but two sequential body plans. This transformation is known as metamorphosis (literally meaning "to change one's shape").

Strange as this story is, metamorphosis is not limited to insects. In fact, animals whose life histories involve two or more distinct body plans are the rule rather than the exception.[5] Most marine invertebrates (animals without backbones) have a metamorphic life history, with one or more free-swimming larval stages, followed by a bottom-dwelling, reproductive adult stage. Parasitic organisms can also have two or more developmental body plans, each specially adapted for a particular host. Even some vertebrates, including frogs, toads and salamanders, go from an aquatic larval stage (tadpole) to a terrestrial adult. Given the widespread presence of metamorphosis as a developmental strategy, one would expect there to be good explanations for its evolutionary origin. Yet it remains an enigma, for a number of reasons.

## I. What the Fossils (Don't) Show

According to standard geological

---

[5] Wallace Arthur, *The Origin of Animal Body Plans: A Study in Evolutionary Developmental Biology* (Cambridge, United Kingdom: Cambridge University Press, 1997), p. 271.

models, multicellular marine invertebrates such as crustaceans, worms, mollusks, and echinoderms appeared for the first time during the Cambrian explosion 540 to 520 million years ago. The sudden appearance in the fossil record of animals with such complex, distinct body plans has spawned many theories but no satisfactory answers. Darwin himself acknowledged that this mystery was the single most difficult challenge to his theory. Even more mysteriously, it appears that the most ancient phyla were metamorphic from the beginning, based on the few larval forms that have been preserved.[6] This suggests that these Cambrian animals had not one but two or more developmental stages at the outset, a small and free-swimming larva, and a bottom-dwelling adult with little or no resemblance to its earlier form. But how such transitions could have evolved, and from what, is completely unknown.

In contrast, insects arrived on the scene much later, in the late Silurian or early Devonian, and apparently developed metamorphosis secondarily. The most ancient insects were wingless, terrestrial animals that developed directly into mini-adults, and lacked any metamorphosis (this is called *ametabolous*—literally, "without changing"—development).

Winged insects such as dragonflies and mayflies appeared in the late Devonian or early Carboniferous. Scientists generally agree that all winged insects came from a single lineage, but debate still rages in the scientific community about how it happened. It appears, based on some fossilized nymphs and adults and from what we know of their modern relatives, that from the beginning these insects had a partial form of metamorphosis (*hemimetabolous*—literally, "part changing"—development). The nymphs resemble adults in many respects, but lack wings and reproductive structures. Through several successive molts their wings grow gradually, with fully developed wings and reproductive organs appearing only in the adult. Other familiar hemimetabolous groups include grasshoppers and crickets.

Insects that undergo complete metamorphosis, such as beetles, flies and ants, did not appear until the late Carboniferous or early Devonian. These insects have been fabulously successful. In fact, nearly 85 percent of all modern insect species have *holometabolous*—literally, "all changing"—development.

Butterflies and moths were among the last to appear on the scene. Their order, the Lepidoptera—literally, "scaly wings"—first appeared in the fossil

---

[6] Richard Strathmann, "Hypotheses on the Origins of Marine Larvae," *Annual Review of Ecology and Systematics* 24 (1993), p. 89.

record in the Jurassic, and more significantly in the Cretaceous. These insects have a dramatic—and well-known—holometabolous life history.

What distinguishes holometabolous species is their strikingly different life stages. The major stages of holometabolous (abbreviated Holo) metamorphosis are (a) egg, (b) larva (often given a different name, such as "caterpillar"), (c) pupa (or chrysalis), and (d) adult, in that sequence:

> **Holo**  egg > larva > pupa > adult

Contrast this with hemimetabolous (abbreviated Hemi) development:

> **Hemi**  egg > nymph > adult

Because hemimetabolous development appears simpler, and because fossil insects with this pattern of development appear earlier in the fossil record, it is thought by most scientists to be evolutionarily primitive, evolving prior to the more complicated pathway seen in most modern insects.

Thus, from an evolutionary standpoint, the problem of the origin of butterfly metamorphosis—in particular, of the pupal stage—is really the problem of the origin of holometabolous metamorphosis generally, not just in Lepidoptera (although the pathway is especially dramatic in butterflies).

Now to outline the scope of the problem.

## II. Directions That Work, Directions That Don't

Imagine being invited to a wedding and receiving with the invitation a hand-drawn map that looks something like this:

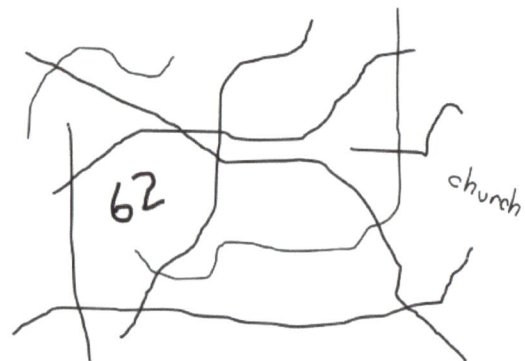

Figure 1. Hand-drawn, impressionistic "map."

The written directions accompanying the map read this way:

- *Take Route 62*
- *Go straight, but not too far*
- *Turn at stoplight*
- *When you see that brick wall, turn again*
- *Church is right there*

About twenty years ago—before GPS devices or Google Maps—my wife and I (Paul) found ourselves in just this situation. (I'm omitting the details to spare the feelings of loved ones.) It's quite certain that the church *was* "right there," somewhere, at least in the mind of the mapmaker—but after nearly an hour of increasingly desperate searching, driving past innumerable candidate stoplights and brick walls, we stopped to ask for help. The locals who saw our map shook their heads in bewilderment. No wonder you're lost, they said.

"Had we but world enough and time," to quote Andrew Marvell, we could have searched the area via an undirected, or random, walk. Sooner or later, the church would have appeared. But, in the finite interval we had, a random walk wouldn't do. And, in practice, no one *ever* reaches their destination using randomly generated directions. That would be known as "getting lost."

In our universal experience, *map directions that work as they should are the product of intelligent design*. The directions describe a pathway linking a starting point—call it A—through a series of specific steps (B, C, D, and so on) to a destination: call it Z. Such directions guide us because of what they *exclude*, out of a vastly greater range of possibilities, which is the very definition of *information*. Hopelessly nonspecific "directions," by contrast—you know that one brick wall? Well, turn there—cannot function to guide any-

thing, because they exclude too little. Such pathways might as well be random walks.

With those contrasting images in mind, let's turn to animal development: in particular, to butterfly metamorphosis, which is development with all the lights flashing, and the siren wailing, too—not to mention a marching band, baton twirlers, colorful floats, and the homecoming queen waving from a convertible.

### III. Animal Development Is a Highly Specific Pathway Across a Magic Bridge

When an egg of a female Monarch butterfly is fertilized by the sperm of a male, that cell, and its many daughter cells, set out on a long, targeted pathway: A to B to C—and so on, to Z, where Z is the adult form capable of reproduction, thus starting the whole cycle again. *The pathway aims at the target of reproductive capability*. Keep that in mind, because we'll come back to it shortly when we consider the logic of natural selection.

But there's another important feature to development—in all animals, not just butterflies—too little grasped, even by many biologists.

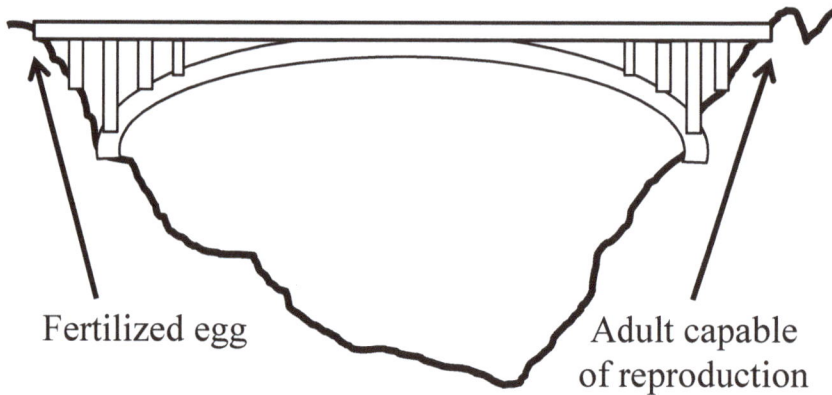

Fertilized egg

Adult capable
of reproduction

Figure 2. The "magic bridge" of animal development, from fertilization to reproductively capable adult.

*"...as long as one keeps walking in the right direction, the bridge will be there beneath one's feet...."*

When an egg divides, the mass of cells to which it gives rise must "keep going" to reach the adult form (defined as being capable of producing gametes, i.e., reproductive cells such as eggs and sperm). One can think about this like someone walking from one side to another across what might be called a *magical bridge*, a structure that would fit perfectly in an Indiana Jones movie.

The bridge is magical (so to speak) because, *as long as one keeps walking in the right direction, the bridge will be there beneath one's feet*. The moment one stops moving, or heads in the wrong direction, the bridge instantly disappears, with dreadful consequences—into the gorge one tumbles.

Developmental biologists know this "magic bridge" aspect to animal development intimately. Early embryos, for instance, are often described by isolating *fate maps* for particular groups of cells. A fate map shows the ultimate or terminal destination, in the adult form, of cell lineages that first arise in the embryo, in positions that often scarcely resemble their final target. But to get to that target, the cells must follow a prescribed path, with unerring trajectories breathtaking in their complexity and precision. *The entire process is required; it can neither stop nor go off track.* An embryo whose development is arrested midway or distorted in a major way will die: the end goal of reproductive adult will be lost.

Butterfly development exhibits these precise pathways, but with the additional aspect of crossing a bridge of astonishing delicacy: namely, the chrysalis. Here, the Indiana Jones magic bridge dimension really does take one's breath away, because during the pupal stage (in the chrysalis), the tissues and structures of the caterpillar are almost entirely dissolved away, digested by cell death processes (known as apoptosis and autophagy) into a molecular soup. The walker on the bridge crosses on a lane just wide enough for each footstep, with a chasm of death on either side—*and the walker must keep moving*. Out of the soup arises the adult form, with its wings, legs, proboscis, genitalia, eyes, antennae, and so forth.

Could this developmental pathway have evolved via the natural selection of randomly arising variation, as posited by neo-Darwinism? To answer that question, we need to look at what the process of natural selection requires—and what it cannot do, in principle.

## IV. Natural Selection: A Real Process, but Entirely without Foresight

The process of natural selection requires three conditions. When these are present in a population of organisms, they are jointly necessary and sufficient for natural selection to occur:

1. **Variation** *in some trait* **p** *(the variations can be molecular, physiological, anatomical, or behavioral—any heritable trait of an organism may be affected).*
2. **Selection** *in relation to the presence or absence of trait* **p***: organisms possessing trait* **p** *must leave more offspring than those without* **p***.*
3. **Heritability** *of trait* **p***: parents must be able to transmit trait* **p** *to their offspring.*

Note that *if any one of these conditions is absent, natural selection cannot happen.* These conditions can be thought of as the three legs of the stool of natural selection.

But there's another aspect to natural selection, which bears critically on the problem of the origin of metamorphosis. Because natural selection depends (with condition 1, random variation) on whatever happens to vary in a species, or not—and there's no way of knowing before the variations occur, or even if they will occur—the process cannot look into the future. Unlike human designers, therefore, who can visualize a distant target, the process of selection "sees" only the variations randomly arising in each generation, and their immediate selective outcomes.

Thus, "life never evolves with foresight," as evolutionary biologists Andrei Rodin, Sergei Rodin, and Eörs Szathmáry explain:

> ...natural selection works strictly "in the present moment," right here and right now, just like a first-aid ambulance—lacking the foresight of potential future advantages....Therefore, just as with any case of step-by-step evolution towards a more complex system, there should be an evolutionary rationale behind each intermediate step.[7]

This aspect of selection places strict limits on what the process can build *de novo*. If a biological system requires multiple independent changes, for instance, no one of which individually confers a selective advantage, natural selection *cannot* be the process by

which that system came to be. That's it: full stop.

Go back for a moment to the opening story, about the badly drawn map and unclear written directions. Let's suppose the sequence of directions were a chain of five independent steps, like this:

$$A > B > C > D > E > Z$$

Here, Z is the church, and A is some major route of entry (say, a superhighway) into the surrounding geographical area. If the hand-drawn map in question is intended to direct guests to the wedding site (and it was), then reaching point D, or even E, won't satisfy the required function. Only the entire sequence, $A > Z$, will count as a success. One needs foresight, or more generally the ability to string together independent stages, from start to finish, to give directions that actually work.

The same is true, only many times over, with animal developmental sequences. A caterpillar-like species would never evolve in the direction of forming a chrysalis, dissolving its vital tissues in the process, unless—somehow—the variations were also occurring, and being preserved by natural selection, which would also enable that species to make it out of the chrysalis stage. And to leave offspring: condition

---

7 Andrei S. Rodin, Eörs Szathmáry, and Sergei N. Rodin, "On the origin of the genetic code and tRNA before translation," *Biology Direct* 6 (2011), p. 14.

3 of natural selection, heritability, requires that variations be transmitted to one's progeny.

But as we inspect the pathway of metamorphosis, what we see is a magical bridge, where literally thousands of independent decisions need to be chained together for the process of transformation as a whole to work. Reproductive capability—one of the necessary conditions of natural selection—lies on the *far side* of the gorge we are crossing. The caterpillar can't leave offspring. Only the adult butterfly can do that.

But to reach the adult, we need the caterpillar, and then we need to dissolve it into a soup—inside a chrysalis where it cannot feed, move, or do much of anything, other than turn into a butterfly.

If one wanted an example of a biological system that could never be explained by natural selection, butterfly metamorphosis would stand at the head of the line.

## V. Surely Evolutionary Hypotheses to Explain the Origin of Metamorphosis Exist?

Yes, they do. We'll sketch them out, and then look at why they don't work.

Traditional adaptationalist explanations for the evolution of holometabolous development include the claim that the differentiation of larval and adult forms allows them to occupy different ecological niches so that they can better exploit food resources and eliminate competition; however, competition for food resources is not well documented in natural insect populations.[8] Another explanation given is that because larvae often feed within plant material as borers or leaf miners or as animal parasites, these life styles could have led to the reduction of wings and other external structures. Yet many larvae are free living.[8] It also has been proposed that metamorphosis allows insects to have more control over their development. Some insects have a shortened life cycle and so can reproduce more rapidly and avoid predation in their most vulnerable stages. Other metamorphic insects can have long life spans, including the Monarch's Methuselah generation.[8] In the end, though, these explanations are not sufficient to account for how metamorphosis came to be. A metamorphic life history may indeed confer all these benefits, but natural selection by definition does not have the foresight to anticipate future advantages. Remember, unless each and every evolving intermediate stage from embryo to adult remained viable, and each and every incremen-

---

[8] David Grimaldi and Michael Engel, *Evolution of the Insects* (Cambridge, United Kingdom: Cambridge University Press, 2005), p. 334.

tal change in developmental strategy made the insects better able to survive and reproduce, a process like complete metamorphosis could not have arisen by unguided, purely Darwinian processes.

What other theories have been proposed to account for the evolution of metamorphosis in insects? The most outrageous of them, the hybridization theory, is not taken seriously by most scientists. It was first proposed by the marine invertebrate zoologist Donald Williamson, based on incongruences he saw among larval and adult forms of various marine animals and their proposed evolutionary histories. He recently extended his argument to include insect metamorphosis in an article published in *PNAS*.[9] The gist of his proposal is that at some point early in evolution, eggs from an organism of one class or phylum were fertilized by sperm from an organism of a completely different type, leading to a new organism with two different, sequentially developing body plans.

This idea has been severely criticized in the literature,[10] and online by Jerry Coyne on his blog[11] and Brendan

Burrell at Scientific American.[12] The criticisms advanced are quite cogent; yet Coyne and Burrell offer no alternate explanation for how metamorphic organisms came to be. The only other viable explanation on the table will be explained below.

When I (Ann) was a graduate student in the 1980s, one of the professors for my oral qualifying exam was James Truman, an expert in insect physiology. At one point in the exam, Truman asked me how I might account for the evolution of complete metamorphosis in insects. I had no idea, having never studied the question. After the exam was over, I asked him what the answer was. He said no one knew. He went on to say that though many theories had been advanced, no conclusive evidence had been found for any theory.

Years later, I learned that Truman must have been actively thinking about the problem of metamorphosis when he asked me that question. In 1999, he and Lynn Riddiford proposed a new hypothesis for the evolution of metamorphosis in insects.[13] They proposed that the caterpillar (or larval) stage in holometabolous insects is homologous to an extended pronymphal stage in hemimetabolous insects.

---

9 Donald I. Williamson, "Caterpillars evolved from onychophorans by hybridogenesis," *PNAS* 106 (2009), pp. 19901-19905.
10 Alessandro Minelli, "The origins of larval forms: what the data indicate, and what they don't," *BioEssays* 32 (2009), pp. 5–8. DOI:10.1002/bies.200900133
11 http://whyevolutionistrue.wordpress.com/2009/09/04/worst-paper-of-the-year/; http://whyevolutionistrue.wordpress.com/2009/10/29/controversal-paper-on-origins-of-caterpillars-debunked/

12 http://www.scientificamerican.com/article.cfm?id=national-academy-as-national-enquirer
13 James Truman and Lynn Riddiford, "The origins of insect metamorphosis," *Nature* 401 (1999), p. 447.

Remember the Hemi pathway for development?

**Hemi** egg > nymph > adult

Now insert a pronymphal stage right around the time of hatching:

**Hemi** egg > pronymph > nymph > adult

In simplified terms, a pronymph is a short, non-feeding stage that in some insects happens right before hatching, and in others right after hatching. This Hemi pathway was hypothesized by Truman and Riddiford to have evolved by gradual steps to Holo, as shown in the diagram below.

> *In this scenario, the development of adult structures, that normally appear gradually during the nymphal stages of a Hemi species, is 'telescoped' into a single pupal stage...*

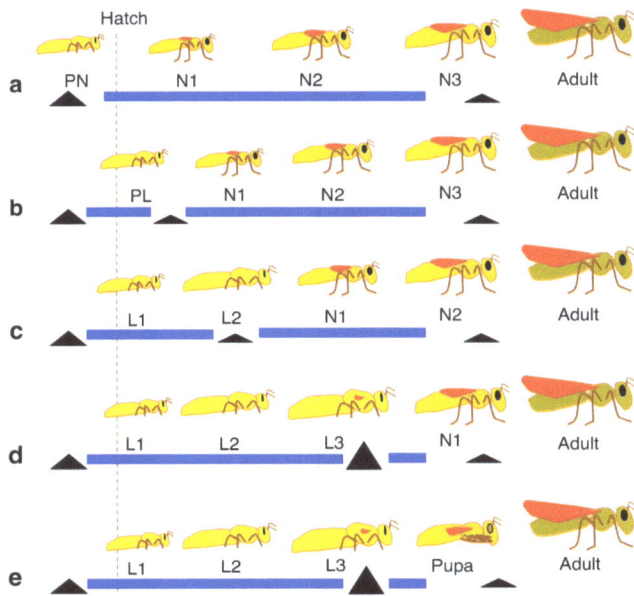

Figure 3. The proposed stages of evolution to complete metamorphosis in insects, adapted from Truman and Riddiford's 1999 paper. JH expression is indicated underneath each sequence by blue bars. The black triangles show where the organism is progressing to the next developmental stage in each sequence. PN, pro-

nymph; PL, protolarva; N1-N3, nymphal instars (only three are shown for simplicity); L1-L3 larval instars.

In this figure, (a) represents the ancestral Hemi pathway. Truman and Riddiford hypothesized that changes in the timing of one of the main hormones of insect development, juvenile hormone, could have caused a larval (extended pronymphal) form to evolve that would be able to feed after hatching (b). Continuing selection for delayed maturation would then extend the larval period (c-d) until a pupal stage evolved (e). In this scenario, the development of adult structures, that appear gradually during the nymphal stages of a Hemi species, is "telescoped" into the pupal stage—giving rise to true holometabolous metamorphosis.

Now, we all have a single experience of the power of hormones: think, for example, of what happens to humans when the hormones of puberty are switched on. Perhaps not surprisingly, it has been found that hormones are important in controlling insect development also. Juvenile hormone (JH) in particular maintains the *status quo* in holometabolous insects. In broad outline, the story is that if JH is constant, the larva molts to another larval stage. When JH is withdrawn in the presence of other hormones, the larva moves to the next stage of development. So by changing the timing of JH expression (blue bars above), Truman and Riddiford argue that the gradual changes of hemimetabolous development could have been converted into holometabolous development.

Thus the caterpillar as a modified pronymph would have to develop the ability to feed and to grow through a series of molts, while suppressing the changes of normal nymphal development. The pupa (or chrysalis) stage would also have to encompass all of adult development, compressed into one tumultuous reorganization rather than stretched over a number of molts.

## VI. Difficulties with This Scenario

Let's remember the logic of natural selection, as it will be our guide in evaluating this evolutionary hypothesis. We can start with the first piece of the story, the origin of the larval form (step [b] in the figure).

In Hemi species, the pronymph (PN, the first stage of [a]) is a non-feeding stage. Thus, if the development of the pronymph is going to be prolonged, to allow it to become a larva, it must simultaneously evolve the ability to feed. "The ability of this stage [PN] to feed would seem to be an essential preadaptation for it to evolve into the larva."[14]

---

14 Ibid., p. 451.

The logic of natural selection asks is there any evidence for such variations—the *de novo* origin of feeding behavior—in the pronymphs of any Hemi species? The answer is no. In hemimetabolous insects, pronymphs may or may not hatch from the egg. Those that hatch are capable of limited movement, but they get their nutrition from remaining egg yolk. Not until the molt to the first nymphal stage do they acquire the ability to feed. So this evolutionary story requires a new kind of pronymph, one capable of feeding. Otherwise, remaining a pronymph is a death sentence, not an advantage.

Thus requirement 1, evidence of relevant variation, has not been met.

A second, related problem is that if you expose a hemimetabolous embryo to high JH, you don't get holometabolous development, you get a mess.[15] Such embryos develop inappropriate nymphal rather than pronymphal characters and/or terminate development prematurely, depending on the timing of the hormone application. This is because these hormones act by regulating a complex network of particular downstream genes. As a result, messing with the timing or level of expression of a hormone is in general detrimental or lethal, not transformative.

In order for any changes to the timing and level of hormone expression to have extended the pronymphal stage into multiple larval stages, there would have to have been an unknown number of regulatory changes and compensatory mutations to make that possible.

But the biggest problems come because delaying development in the caterpillar requires that all of adult development be compressed into a single pupal stage. This is the point of transition between steps (d) and (e). In Holo species, there is considerable dissolution of cells and tissues during the pupal stage and adult structures are constructed during this phase.

For this to work, there must be some way to set aside cells that will build the adult tissues (wings, eyes, antennae, etc.) and to trigger their development into the right structures at the appropriate time. How this is done varies among different types of tissues. For example, some caterpillars reuse leg epidermal cells to contribute to the development of adult legs; this means that these cells must be able to produce two different kinds of cuticle. But caterpillars also use cells (called imaginal primordia or disks) that were set aside in the embryo just for making adult structures; these cells may remain quiet in the caterpillar and only begin to grow and produce adult struc-

---

[15] DF Erezyilmaz, LM Riddiford, and JM Truman, "Juvenile hormone acts at embryonic molts and induces the nymphal cuticle in the direct-developing cricket," *Development Genes and Evolution* 214 (2004): 313.

tures in the final larval stage and in the pupa. How do cells acquire the ability to carry out two different, sequential programs in response to hormonal signals? How do cells get set aside for future use? And how do you coordinate the two sets of cells to make fully functional adult structures?

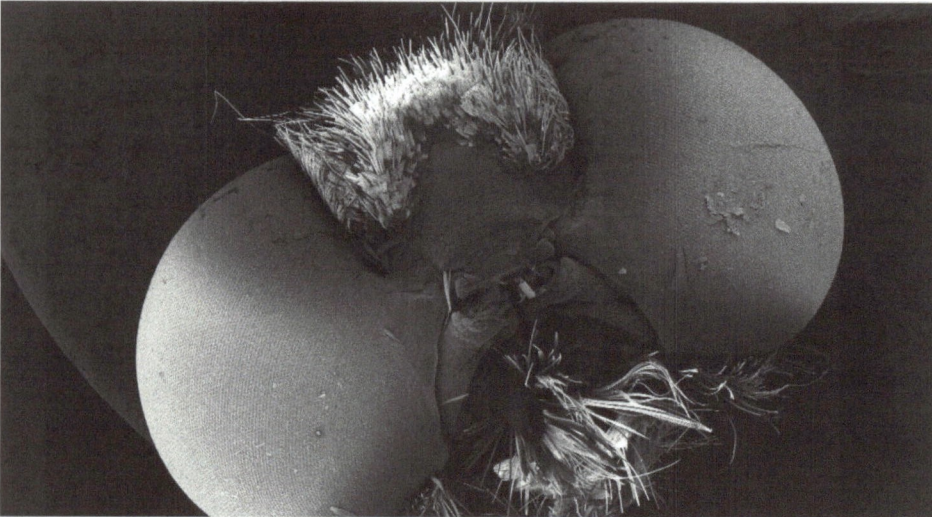

*" The flip side of the pupal riddle is that there also must be a way to specify and coordinate which larval cells will self-destruct. "*

The flip side of the pupal riddle is that there also must be a way to specify and coordinate which larval cells will self-destruct. Programmed cell death is common enough in development as a way of remodeling tissues into their final form (how natural selection can produce a process like programmed cell death is another question), but only in metamorphosis is programmed cell death so wide-spread and catastrophic. Insects with hemimetabolous development change gradually with each molt and so do not require extensive remodeling, but insects with holometabolous development remake themselves in one compressed stage. In response to the hormonal cues that trigger pupation, the majority of caterpillar cells initiate a cellular program that leads to their destruction—essentially, they commit suicide in order to make room for the new adult structures.

This represents a new response to hormonal signals at the time of pupation. But new responses to hormones require new sets of receptors and genetic response elements. In order for programmed cell death to take place where it should, and nowhere else, these elements would have to be present *in advance* in the appropriate larval tissues.

Presumably, then, the response elements responsible for turning on the cell death program in the appropriate tissues would need to be in place, but not yet activated, by step (d) of the evolutionary scenario. But how do you select for something not yet active?

*And unless simultaneous mutations occur to enable the building of adult structures after or during tissue and cell dissolution*—between the pupal and adult stages in step (e)—requirement 3 of the logic of natural selection, heritability, will not be met. The start of the pupal stage would be a death trap.

That's a lot of serendipitous change to produce the pupal transition. But there's more: complex behaviors are also required for the process to go right. Getting any part of this wrong would be disastrous for the organism. With its tiny insect brain the caterpillar must know to stop feeding, climb up away from its food, and find a safe place to pupate. If appropriate to its species, it must spin itself a cocoon. Then it sheds its caterpillar skin, dissolves its larval tissues (but only the right ones), and triggers the adult cells to finish dividing and differentiate. These cells knit everything together, make new cuticle, and the nervous system makes connections to the new muscles, eyes, and antennae. When all is complete, and the environmental cues are right, the caterpillar's brain triggers eclosion. It emerges from the pupa, pumps fluid into those gorgeous wings, hangs there until the wings harden and its proboscis is knit together, and then it takes off. Suddenly it sees and smells things it has never seen or smelled before. The world has become a new place.

Finally, it may be that the claim that the pronymph is analogous (or even homologous) to the caterpillar, and the nymph to the chrysalis, is valid in a distant sort of sense. There are intriguing similarities in some aspects of nervous system development and cuticle formation in both. But to make the claim that the stages are similar is to overlook the radical differences that also exist. The bottom line is, even if the caterpillar is a modified form of the pronymph, the modification of pronymph into caterpillar is unlikely to have come from an unguided process. There is also no gradual step-wise process that could create something like the pupa to adult transition. Purely Darwinian processes (mutation, natural selection, and genetic drift) lack the foresight to have created such a process.

## VII. Conclusion: Trust Your Intuitions on This One

Intuition is not always a reliable guide. Many people, on seeing (for instance) a spiraling, descending wooden ramp carrying a ball bearing, think that the ball bearing will continue rolling off the ramp in a curving trajectory. Those with a bit of basic physics in their backgrounds, however, know that the ball bearing will come off the ramp in a straight line. Intuition needs to be overruled by evidence.

With butterfly metamorphosis, however, all the evidence we have reliably supports the intuition that this complex sequence could not have evolved by an undirected process, such as natural selection. Indeed, it is the very logic of natural selection itself that tells us the process won't work. The theory makes strict evidential demands on investigators, and unless those demands are met—and, in the case of butterfly evolution, they haven't been—we are simply not entitled to deploy natural selection in explanation.

More to the point, all the evidence points towards the reality of intelligent design. In our ordinary experience, systems requiring multiple independent components, aiming at and hitting a distant target, and the re-use of lower-level modules for very different functional roles, implicate a designer, with a mind.

Or a Mind, where the capital letter indicates—something special.

This is a rational inference, from evidence. It's science in the old-fashioned sense of *knowledge*. You can count on it.

*ysippus* ♂

*ysippus* ♀

*s f. dorippus* ♂

*eptentrionis* ♂

*mniace* ♂

*H. misippus* ♂

*H. misippus* ♀

*P. ceylanica* ♀

*Ch. clytia* ♂

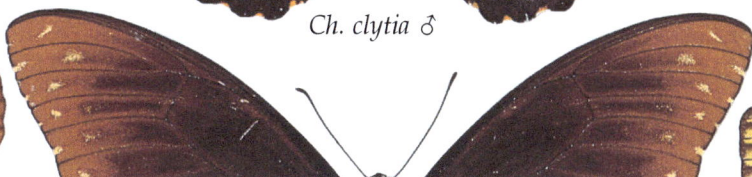

*E. hypermne*

*E. hyperbiu*

*E. hyperbius*

*P. ceylanica*

*D. ceucharis* ♀

# Chapter 5
## Mimicry and Protective Resemblance:
## A Philosophical Appreciation
Bernard d'Abrera

# Chapter 5
## Mimicry and Protective Resemblance: A Philosophical Appreciation[16]
Bernard d'Abrera

Butterflies are created natural things that can be recognized as existing in time, in thousands of different kinds. Within each kind, butterflies survive for short periods of time as individuals, and for longer periods of time as species. Their continuing existence as species over time depends on the survival of the individuals. Within created sentient nature, the individuals are programmed to do the best within their mental and physical capacity, in order to survive and ensure the continuation in time of their kind. There are many hazards to this survival. Indeed nature, in its entropic condition, is pathologically beset with hazards, and while the hazards are not entirely the product of blind chance, they are nevertheless predictable by the creature to a lesser or greater extent. For instance, a butterfly is aware of sudden movement, or extremes of temperature, or the change in the intensity of light, and can by instinct (programmed, it would seem, by an external intelligence), and by experience (learned through its own intelligence), take steps to minimize or escape the consequences of such hazards during its individual lifetime.

Thus butterflies are by no means exempt from the predatory hazards that beset most of nature. Indeed (with one notable exception) they are among the weakest and least aggressive of all creatures, totally lacking the active physical means to exert force in their own defense. All they have at their disposal (as individuals) is the means of swift escape in flight or (as species) the sheer weight of numbers. But there are many species that are either not possessed of the ability for swift flight, or not programmed with the genetic propensity to reproduce themselves in huge numbers. Thus it is that such *actively* defenseless creatures have been endowed with effective *passive* means for the survival of their kind.

The passive means of survival of many creatures involves a genetically programmed capacity for deception or fraud by species involuntarily expressed in the morphology of its individuals. This fraud may belong to one or other of two types, Protective Resemblance and Mimicry. One of the most celebrated examples of Protective Resemblance in the avian world is that of the Frogmouth (*Batrachostomos* sp.) in which the bird at rest on the forest floor or upon its nest resembles a

16 Reprinted with permission from Bernard d'Abrera, *The Concise Atlas of Butterflies of the World* (Melbourne, Australia: Hill House Publishers, 2001).

decaying log. Its camouflage colors and markings and the way its head hunches into its body, all maintained rock still, ensure that it is totally missed by predators' eyes hungry for a meal of Frogmouth or Frogmouth's eggs. There are other examples of Protective Resemblance throughout the bird world, especially in the cryptic coloration of many female waterfowl, ducks and so on, but it is in the insect world in particular that this phenomenon reaches its most spectacular. There must be few people indeed who have never heard of or seen stick—or leaf—insects (Phasmidae), butterfly or moth caterpillars that resemble twigs (Geometridae), and so forth; but in butterflies, we have larvae that pass for dried bamboo leaves, snakes, or fresh bird droppings, pupae that cannot be distinguished from dried twigs, and actual butterflies (*Kallima* sp.) that are so convincingly leaf-like on their verso surfaces that no two individuals (even from the same brood) are quite alike. These *Kallima* species have wing shapes that are dried-leaf-like, complete with stem and leaf tip, midribs and veins, various mottling due to age and decay, and even the little bits of leaf mold or fungus that one might expect to find on a dried leaf.

The most astounding and famous case of Protective Resemblance amongst lepidopterous insects was reputedly that of the White Pepper Moth (*Biston betularia*) in Europe or North America. The story went something like this. These moths were so colored that when they rested on the trunks of lichen-covered trees they "vanished" from sight. With the advent of the Industrial Revolution, factories gushed out huge quantities of pollution, which caused the lichen cover to disappear, and the trees to darken. The white spotted moths very soon lost their "protective resemblance" and became fatally visible to predatory birds. But within a miraculously short period of time a dark colored form (f. *carbonaria*!) emerged and the moth once again vanished from view as it rested on the now blackened tree trunks. Julian Huxley triumphantly announced that here was "Evolution before our eyes."

Unfortunately, all of the foregoing relating to the adaptability of *Biston betularia* to a rapidly changing environment was later exposed to having been a colossal fraud. All the experiments intended to demonstrate the evolutionary aspects of these changes were conducted by one Bernard Kettlewell. What Dr. Kettlewell had in fact done was an elaborate but somewhat amateurish set-up in order to demonstrate the different phenotypes contrasting with non-homogeneous backgrounds. He had deliberately placed long-dead, museum-set specimens in entirely artificial posture on tree-trunks for his photographs, some of these specimens even having pin marks on their thoraxes! Lepidopterists have long known that *B. betularia* has always had an enormous range of infrasubspecific phenotypes, ranging from almost pure white to almost totally charcoal in ground-color. They would also be seen resting on any background, whether complementary or not. There is no empirical evidence that the different forms would deliberately choose to rest against backgrounds that would allow them to "vanish."

On the other hand, the true scientist must know that phenotypic variation is a capacity inherent in every population, and that profound environmental shifts (or "shifting selection") can cause changes in genotypic frequency until a new equilibrium can be reached. The process is, by inference, bi-directional, and in some cases even multi-directional, but at no time has the species (or kind) actually and permanently changed into a new species. This type of observable (and demonstrable) natural selection is most repugnant to the evolutionist psyche, because notwithstanding all the enormous variations, there is no permanent mutation into something genetically new; nothing in fact changes. *Plus ça change plus c'est la même chose!*

The other type of deception or fraud in nature is referred to under the heading of Mimicry. The most useful way to describe this phenomenon is to refer to the plate of Ceylon butterflies figured here and to compare the specimens illustrated.

The three butterflies in the top left hand column represent both sexes of the species *Danaus chrysippus* (family *Danaidae*). This is a poisonous species. It is poisonous not because it goes around biting things and killing them, but because it belongs to a family whose larvae, having eaten poisonous food plants, are themselves full of the poisonous alkaloids of those plants (of the botanical families *Asclepiadaceae* or *Apocynaceae*).

Thus if a putative predator were to ingest the adult butterfly, it would become very sick indeed. Now, the theory goes that because of their poisonous nature, and by dint of experience, the predators (who themselves have to fulfill their own survival program) learn to recognize the markings and colorations of the different poisonous species and leave them alone. The theory further goes, that certain non-poisonous butterfly species (the mimics, whose bodies do not contain toxins) resort to passing themselves off as close copies of the poisonous species (the models).

The middle column comprises both sexes of the non-toxic species, the Danaid Eggfly (*Hypolimnas misippus*, family *Nymphalidae*). The first four butterflies on the right hand column represent both sexes of two other non-toxic species, the Palmfly (*Elymnias hypermnestra*, family Satyridae) and the Fritillary (*Argynnis hyperbius*, family *Nymphalidae*).

It will be noticed that the female specimens in each column all resemble each other superficially in the general orange coloration with black f.w. apex and white transverse sub-apical band. This resemblance becomes all the more noticeable when comparing their respective males. One would expect the females at least to resemble the males, but they do not. Instead they resemble both sexes of the toxic model.

Now the plot thickens somewhat when the third specimens in each of the first and middle rows are compared with the black-tipped f.w. males above them. Evolutionary mimicry enthusiasts immediately point to this and say that it is a fine example of a species evolving on the basis of the survival of those individuals most able and suited (the fittest) to do so. "So clever" (they say) is the deception by mimicry that even when the model (the poisonous species) has a different form of the female, the mimic goes so far as to copy that as well. This different form is the so-called f. *dorippus*. One will notice however that the other two mimics in the right column (*Elymnias* and *Argynnis*) do not have an equivalent *dorippus* form. Nor do the males of all three non-poisonous species figured here show the slightest tendency to change their appearance. Why not?

Other examples of mimicry on the same plate are quite self-evident; the three lower specimens in the left column are all toxic danaids; the three lower specimens in the middle row (in parallel with them) are a non-toxic pierid (totally dissimilar blue male in

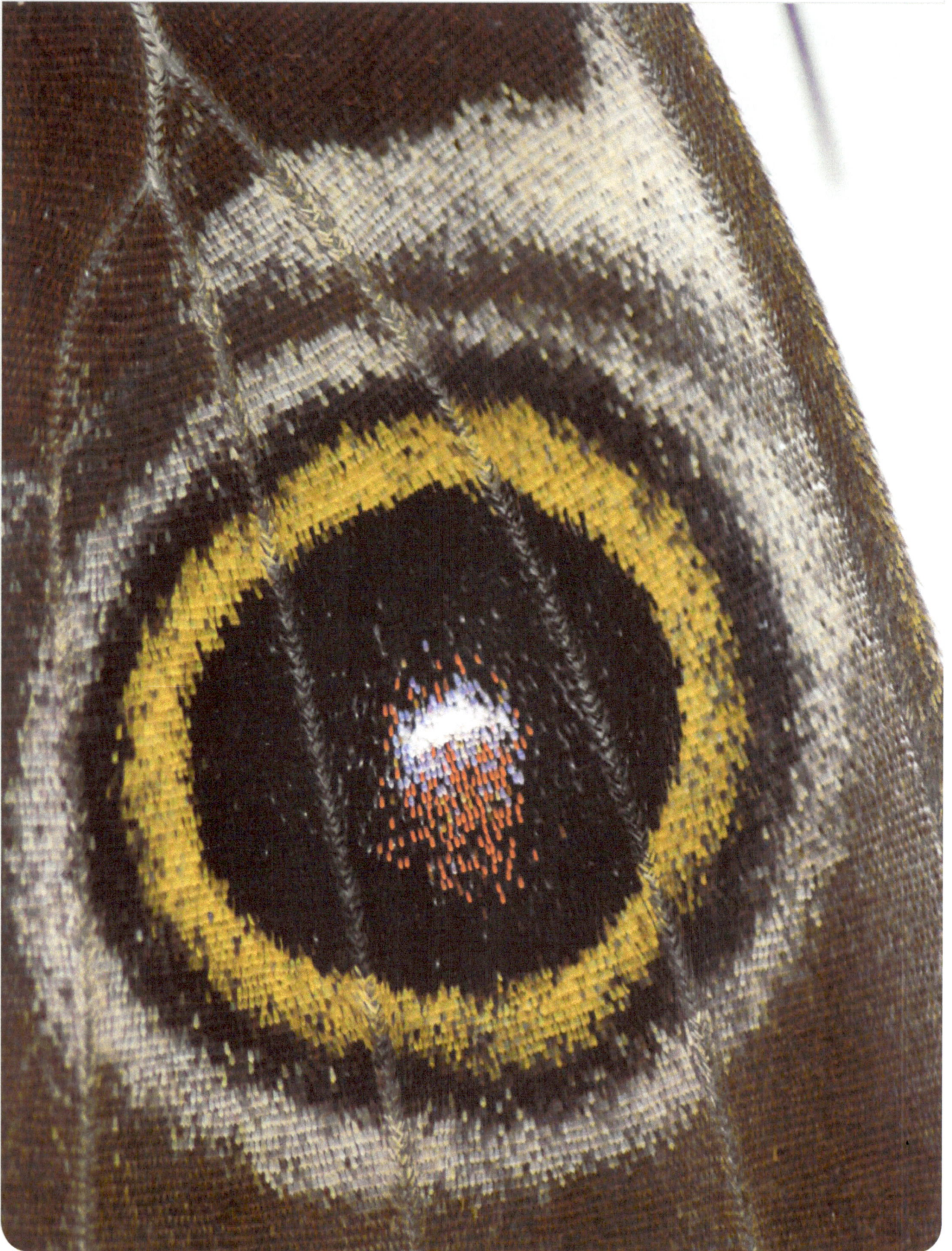

the right hand column), and both forms of the non-toxic Mime Butterfly (*Chilasa* sp.) which allegedly mimics the toxic danaids.

The last two specimens in the right hand column are both pierids. The upper one, the Jezebel (*Delias* sp.), has some toxicity because its larvae feed on the Mistletoe (*Loranthus* sp.)—the lower one, the Painted Sawtooth (*Prioneris* sp.) is apparently non-toxic, because its larval food plant is the Caper (*Capparis* sp.) The bright colors of the model warn prospective predators of its toxicity. The imitative bright colors of the mimic exploit this warning for its own benefit.

It must be stressed that all of this fraud and deception is not voluntarily willed by the creatures. They cannot and do not choose to appear thus, but they certainly have been programmed within their specific kind to appear so. Hence morphology is a composite set of particular structural attributes or diagnostic features strictly and invariably peculiar to a given species. Stated simply, the capacity of deception build into *Kallima* butterflies cannot be expressed at any point in time by, say, Swordtail or Swallowtail butterflies. Thus every species has *locked* into its genome its singular and strictly *limited* capacity to reproduce only *its own kind* or cease to exist (survive) as a species. Because

of its inbuilt and secure genetic programming, it *cannot* change (mutate) into another species. But while it can have within that program the genetic capacity to produce individuals that permanently resemble other species (to human eyes at least), leaves, twigs, snakes, birds (hummingbirds, hawk moths), bird droppings, monkeys' faces and so on, it does not mean that it can change (genetically) into those kinds of things!

There are two principal kinds of mimicry, Batesian and Mullerian, but it hardly matters what they are called, because the point is that if the butterflies are not personally responsible for acquiring the capacity for protective resemblance or mimicry (and we are all agreed that it is a neat (intelligent) trick), then who or what intelligence put it there? For it to happen in a single species once through chance, is mathematically highly improbable. But when it occurs so often, in so many species, and we are expected to apply mathematical probability yet again, then either mathematics is a useless tool, or we are being criminally blind.

*Thus,*

(a) any given specific *kind* differs (*inter alia*) from any other specific *kind* by the uniqueness of its genetic make-up (usually demonstrable by qualifying chromosome type and quantifying its number);

**(b)** genetic information by any individual within a specific *kind* can only be transmitted (naturally) by inheritance;

**(c)** the only source of such inheritance in nature is from sexually complementary parents, and finally from one original male and one original female of that specific *kind*;

**(d)** the original male and original female must have had sufficient gene vigor to sustain millions of generations before entropy finally brings their line to non-viability. Therefore such a male and such a female must necessarily have been superior to their progeny, and could only have been created fully programmed, "ready fashioned," with all the genetic information required for the survival through time of their specific kind. This survival would also necessarily encompass any variations imposed upon the population that would permit the species or kind to adapt to changes in environment or other pressures. For example, such extrinsic dynamics as changes in climate and elevation or isolation through geological upheav-

al would be the primary cause of race formation. It must be stressed that evolutionists erroneously refer to such changes as "microevolution." They are nothing of the kind. I repeat, they are simply a built-in or programmed response of a species or kind to extrinsic dynamic change. The proof of this is that when isolated populations (races) with clear morphological differences from other related isolated populations are allowed to mix with their relatives, they usually disappear as individual races, returning to the species or kind, another example of so-called "evolution in reverse." Indeed, it is well proven that if many of these isolated populations remain cut off for very long periods of time, the gene pool diminishes in its vigor, and the population perishes. (Dog, cat, and avian breeders know this sorry state of affairs all too well!)

**(e)** Evolutionists refuse to see this, and in fact propose quite the opposite. In several of my previous works, I quote the words of a great geneticist, Professor Maciej Giertych, on this subject.

**(f)** In all species of butterflies the sexes are morphologically different to a greater or lesser extent. This is called sexual dimorphism. To the scientist, whose ontological sensibilities are unencumbered by fashionable skepticism, and who thus understands intelligent, ordered creation, there is no need to explain the obvious. But for the evolutionist, who bases everything on unintelligent and unintelligible chance, there is a need to explain in logical terms (in other words, by the rules of philosophy) the following:

At what point in the evolution of these species did the males and their respective females, both emerge in time, so as to

a) Recognize each other as being of the same species, in spite of their differences?

b) Get together to mate successfully, to reproduce their own kind—because just one evolutionary state of error in trial would guarantee extinction?

In other words, if either sex of each species descended from a common ancestor, was the evolution of the form of each sex (in other words its "morphism") absolutely parallel throughout evolutionary history, so that at each stage of the evolution of that species, blind, unintelligent chance would determine that they continue to express different morphologies, but still maintain specific homogeneity and thus be able to mate successfully?

The evolutionary model requires much faith. Hence the need for evolutionists to face the inherent self-contradiction in their belief in continual ordered existence (through reproduction) of millions of specific kinds of living creatures (arising together, in time), entirely from nowhere, by disordered, materially non-existent chance.

Evolutionism (with its two eldest daughters, phylogenetics and cladistics) is the only systematic synthesis in the history of the universe that proposes an Effect without a Final Cause. It is a great fraud, and cannot be taken seriously because it outrageously attempts to defend the philosophically indefensible.

Some readers may well ask what all this has to do with the simple study of butterflies. My response is that the study of butterflies is not simple, and what I have to say has everything to do with it. I have written this chapter because I want science and scientists to become philosophically accountable.

Evolutionists are notorious for two things. They are the masters of wooly thinking, and are totally incapable of logic in the classical sense. Such woolly thinking reaches even greater clouds of nebulosity with pronouncements such as, "One has only to wait: time itself performs the miracles...Given so much time, the 'impossible' becomes possible, the possible probable, and the probable virtually certain."[17] And this gobbledygook from a Nobel Laureate, Harvard biologist George Wald (1906-1997)! His statement is not only philosophically bankrupt, it is scientifically irresponsible. Its tendentious construction, poor logic and tone of childish credulity make it sound like an advertisement for Disneyland, or worse, like a passage from a second-rate fairytale (for grown-ups).

Natural Science is now in grave disrepute. It survives in its present form only because of a media- and academia-generated program of propaganda which needs the constant distractions of novelties, spurious discoveries, outright fraud, and smoke-screens of personal invective, all of which are designed to keep the punters guessing, and ordinary people from asking the most fundamental of philosophical questions about cause and effect, reason and purpose, and loss and gain. The cruelest trick perpetrated on the most sentient of terrestrial creatures is that they have been cut off from all knowledge of the very fount of their own existence. This is not science. This is unmitigated wickedness.

The true scientific model of the origin of species, by design not by chance, can be demonstrated by common sense and the simplest of scientific experiments with any species, particularly of butterflies; whereas it is scientifically impossible to sustain the evolutionary model without having to distort objective truth itself. However, in this unfortunate age such a notion of truth has come to be denied near universally, precisely because of the damage done by the opposite notion of constant change (or evolution). Thus almost everyone now believes to a greater or lesser extent that truth is a function of time, instead of the axiom that time is a finite dimension at the service of infinite and unchangeable Truth.

---

[17] George Wald, "The Origin of Life" in *The Physics and Chemistry of Life: A Scientific American Book* (New York: Simon & Schuster, 1955), p. 120.

# Chapter 6

## What Is It About Butterflies that Drives Men to Doubt Darwin? Bernard d'Abrera, with a Note on His Curious Encounter with the Smithsonian Institution

David Klinghoffer

## Chapter 6

# What Is It About Butterflies that Drives Men to Doubt Darwin? Bernard d'Abrera, with a Note on His Curious Encounter with the Smithsonian Institution

David Klinghoffer

In Chapter 7 we will meet novelist and lepidopterist Vladimir Nabokov, a self-described "furious" critic of Darwinian theory. An erstwhile butterfly researcher and curator at Harvard and the American Museum of Natural History, Nabokov thought that butterflies possess powers of mimicry inexplicable on Darwinian assumptions. In the same tradition of butterfly-induced Darwin heresy, allow me to introduce Bernard d'Abera, who contributed the essay immediately previous to this one (Chapter 5).

A kind of latter-day Audubon of Lepidoptera, d'Abrera is a philosopher of science, renowned butterfly photographer, one of the world's most formidable lepidopterists—and if anything, an even more furious Darwin doubter than Nabokov. His series of enormous volumes, *The Butterflies of the World*, a heroic act of categorization and illustration, is almost completed with the recent publication of *Butterflies of the Afrotropical Region, Part III: Lycae-*nidae, Riodinidae, in a revised edition including a lengthy assemblage of introductory essays. The latter comprise one of the most colorful, amusing, enraged, and wildly unclassifiable attacks on Darwinism that I've come across.

The book is huge—I've been carrying it around as I bicycle to work and my sore back attests to this—and gorgeously furnished in the systematic section with d'Abrera's incredibly detailed butterfly photos. His pictures were taken both in the field and in the unsurpassed collections of the British Museum (Natural History) where he has been a longtime visiting scholar in the Entomology Department. Unfortunately, priced at more than $500 a copy, the book probably isn't a realistic purchase for you unless you have a professional or at least very serious amateur interest in butterfly classification.

D'Abrera is an old-fashioned scholar, insisting over and over on the indispensability of Linnaean taxonomy, governed by "the rules of ideology-free, empirical science," before it began to be overtaken by worldview-driven speculation. He recalls his dismay in the mid 1980s when his British Museum colleagues permitted themselves to be swept away by a "sudden and almost manic drive to abandon the vestiges of tradition and endow their own output with the pseudo-intellectual flavor of phylogenetics." Candidly, I should say that Bernard d'Abrera, while a distinguished scientist, publisher and photographic artist, takes some intellectual paths for which one cannot commend him. Yet his objections to Darwinism are illuminating, and more fundamental than Nabokov's thoughts on mimicry.

He pours particular scorn on the late Harvard zoologist and would be Darwin heir Ernest Mayr, from whom d'Abrera offers a quotation that sums up everything he finds fraudulent in evolutionary thinking. Mayr explained how evolutionary biology's status as a "historical science" exempts evolutionists from normal standards of scientific argumentation:

...the evolutionist attempts to explain events and *processes that have already taken place.* Laws and experiments are inappropriate techniques for the explication of such events and processes. Instead one constructs a historical narrative, consisting of a tentative reconstruction of the particular scenario that lead to the events one is trying to explain [emphasis added].

The evolutionist begins with the assumption that the events in question *have already taken place, life's development has occurred, by means of Darwinian processes.* He seeks only to "explicate" in more detail how this happened. His method consists of imagining a historical scenario and then spinning out a fictional narrative, in line with a theory that's already held to be true before any proof has been offered. When you reason this way, as Marxists and Freudians also found in their respective pseudo-scientific fields, it's almost eerie how all the evidence you consider appears to uniformly confirm your theory.

This is d'Abrera's basic complaint about Darwinism. His case, however, would not be completely depicted without referring to his interesting recent experience with the Smithsonian Institution, a wonderful taxpayer-supported educational establishment that has a bad record when it comes to treating scientific Darwin-doubters with due respect for academic freedom and free speech. Now to this list of indictments add respect for intellectual property.

Readers may recall the Richard Sternberg affair, in which supervisors at the Smithsonian's National Museum of Natural History (NMNH) persecuted an evolutionary biologist on their staff just for editing a peer-reviewed research paper supportive of intelligent design. More recently, senior figures at the Smithsonian may have pressured the affiliated California Science Center to cancel a contract to show a Darwin-critical documentary, in what seems to be an instance of a public facility illegally regulating speech.

In both of those cases, the indications suggest it was the intention to squash a controversial viewpoint that motivated Smithsonian personnel. In the case of Bernard d'Abrera, there's no reason to believe that it was his Darwin-doubting itself that led to an act of startling brazenness.

Brazen what? "Theft," as d'Abrera calls it in his account published now for posterity in *Butterflies of the World*. He actually puts the word in quote marks since, he observes wryly, his attorney advised him that while it looks to the untrained eye exactly like theft, it wasn't a criminal case, ending up instead in the Court of Federal Claims.

D'Abrera tells the whole story in the aforementioned *Butterflies of the Afrotropical Region, Part III*. A Smithsonian publication on the butterflies of Myanmar (a/k/a Burma) pirated—or "pirated"—a huge batch of his famous, gorgeous, and unique butterfly photos for use without permission, notification, or compensation of any kind. When caught, no one with the Smithsonian tried to deny it. D'Abrera received no apology. The Smithsonian strenuously resisted attempts to claim compensation but, after wearing d'Abrera down, finally agreed to settle the case with a payment of $120,000.

When I contacted several Smithsonian spokesmen, I was referred to the institution's Associate General Counsel, Lauryn Guttenplan, who emailed me that the Smithsonian confirms the settlement was reached but has "no comment on Mr. d'Abrera's version of the dispute as set forth in his new book."

Isn't it strange for a government entity to be so unforthcoming with an account of its actions? It gets stranger. D'Abrera records that in email correspondence with Stephen Kinyon, compiler of the book in question, *An Illustrated Checklist for the Butterflies of Myanmar*, Kinyon was open about having appropriated the lepidopterist's images. He wrote that he "did scan a number of butterfly pictures from your *Butterflies of the Oriental Region* Parts 1, 2 and 3." What number, exactly? D'Abrera counts 1,352 butterfly pictures, or 62 percent of the images "compiled" by Kinyon. That is a lot of butterflies.

Kinyon went on to say that the project was funded by the Smithsonian's Conservation and Research Center (CRC) to help train Burmese forest rangers. The book also got money from the Walt Disney Company Foundation.

Kinyon explained that:

...when putting the checklist together, we realized that the specimens I collected...would not come near to covering all Burmese species. We filled in the gaps with images copied from several published works. I discussed this with my friends at the CRC, and decided eventually that this should not be a deterrent, given the charitable purpose and distribution limited to NWCD [Myanmar's Nature and Wildlife Conservation Division] and the University of Yangon in Burma.

Since when did seeking to educate students of forestry or any other field exempt you from paying someone for his work? On contacting the Entomology Department at the Smithsonian's National Museum of Natural History for a reaction, d'Abrera received from spokesman Gary Hevel the astonishing consolation that curator of Lepidoptera Bob Robbins "comments that Kinyon's use of some of your images is a true testament to your photographic skills."

This is like someone surreptitiously lifting the wallet out of your pocket, extracting the cash inside and making off with it—and then, when confronted, explaining that he talked it over with friends beforehand and decided that since the money in his own wallet "would not come near to" covering an expenditure on behalf of a deserving Burmese acquaintance, therefore the fact that it's your money "should not be a deterrent" to pickpocketing you. On the contrary, you should feel flattered since the choice of you above other possible victims is "a true testament" to your "skills" in earning the money in the first place.

The only problem with this analogy is that there were no other possible victims with an adequate holding of butterfly photos to swipe. In the world of Lepidoptera, d'Abrera is without a competitor. "He is a controversial biologist," comments Dr. Thomas Emmel, director of the Florida Museum of Natural History's McGuire Center for Lepidoptera and Biodiversity, who also appears in the documentary film *Meta-morphosis*, "but one whose remarkable lifetime accomplishments publishing an illustrated catalogue of butterflies of the world must be admired for a unique contribution that will likely never be duplicated."

D'Abrera claimed he was owed $1 million in statutory damages, both for the swiped photos and the accompanying systematics with all the original research that the latter represent. It's not hard to make the case for this figure. D'Abrera was allowed to photograph the enormous and unequaled collections of the British Museum (Natural History), a privilege he paid for, with the whole project consuming six years of his professional life and, he says, an investment of his money equal to a seven-figure sum.

As d'Abrera informs me, the Department of Justice sought to defend the Smithsonian based on a claim that d'Abrera did not deserve such damages because his works, published in Australia where he lives, were not properly registered for copyright in the U.S. However, while I'm not a lawyer, it would seem that d'Abrera should have been protected anyway under the Berne and Universal Copyright Conventions of which the United States is a signatory.

What does it all mean? This is not another Sternberg case. And NMNH spokesman Randall Kremer denies that the Entomology Department at Sternberg's former institutional home had anything to do with the publication of Kinyon's entomological catalogue. Copyright cases are handled for the most under civil law, and they are, of course, common.

But there is something really unseemly and, as I said, brazen about the Smithsonian's treatment of this distinguished 70-year-old biologist. Unlike a lepidopterist at, say, the Smithsonian Institution, d'Abrera is freelance, meaning that either he earns his keep or he starves.

While the d'Abrera affair doesn't tell us anything about the intellectual freedoms denied to Darwin doubters, it does suggest a possible reason why, in those cases, the name of the Smithsonian Institution keeps coming up. On one hand, the SI is a public endowment, answerable to taxpayers and the U.S. Constitution. That demands a degree of transparency and fair play that would not be legally required of private organizations, which are entitled to exclude viewpoints that owners or administrators dislike.

In the d'Abrera story, the Smithsonian took full advantage of its protections and resources as a government entity. Yet when it suits them, staff at the Smithsonian (and the California Science Center) behave as if they were not employees of a public trust at all but rather of a private agency and, in suppressing dissenters from institutional orthodoxy, a rather opaque and unprincipled one at that.

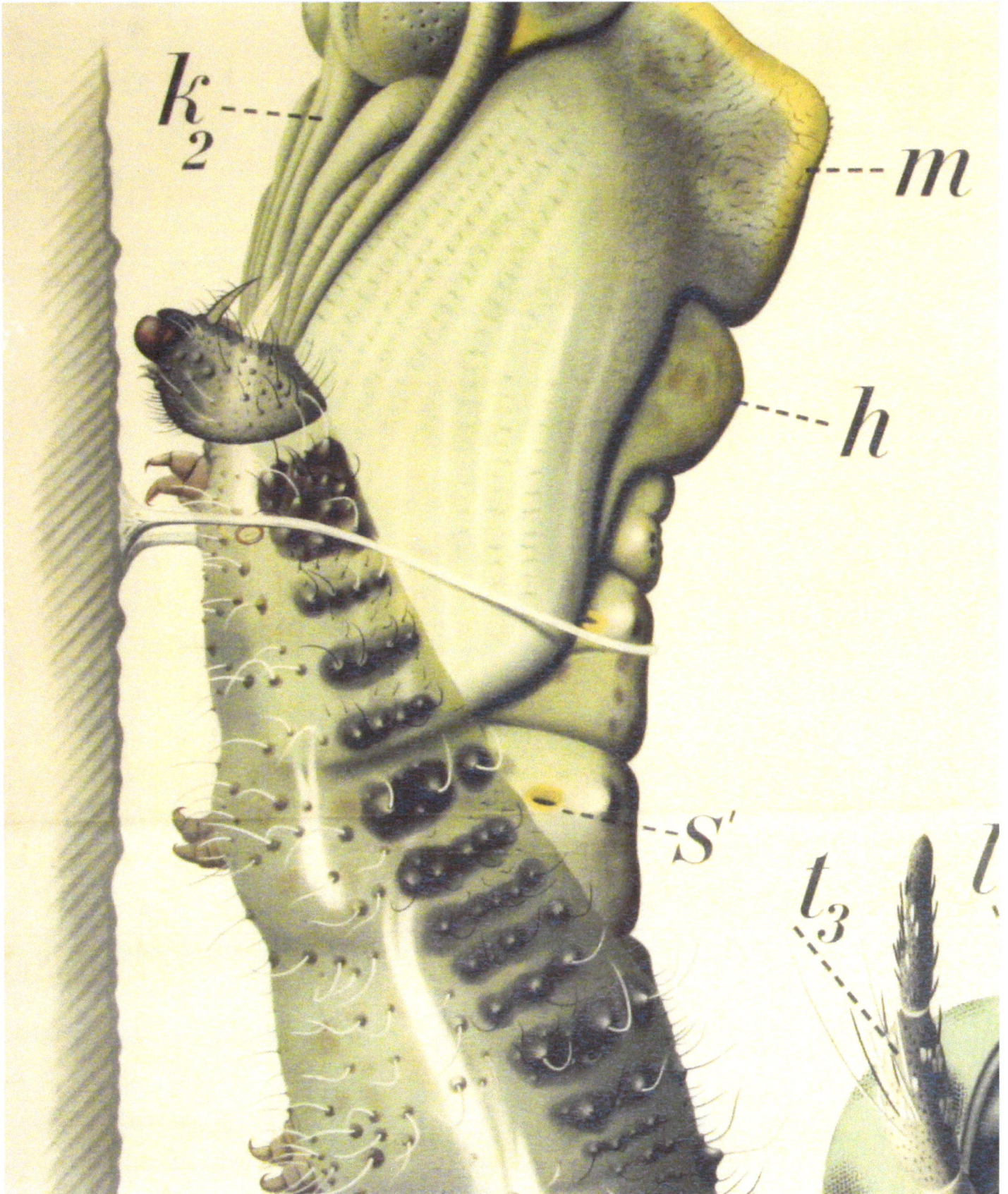

$k_2$

$m$

$h$

$s'$

$t_3$

# Section III: Butterflies and Evolution: History, Science, and Art

# Chapter 7

## "The Grand March of Nature": The Evolution of Alfred Russel Wallace's Intelligent Design

Michael A. Flannery

## Chapter 7

# "The Grand March of Nature": The Evolution of Alfred Russel Wallace's Intelligent Design

Michael A. Flannery

Alfred Russel Wallace knew butterflies. He was first introduced to these most colorful and elaborate of creatures in the tropics of the Amazon River Basin where he amassed, by one estimate, a collection of 10,000 specimens largely consisting of butterflies and beetles. During his eight-year stay in the Malay Archipelago he collected more than 13,000 butterfly specimens alone![18] The complexity of this animal's life cycle, its metamorphosis from a humble and voracious caterpillar to a flying insect of unparalleled beauty, fascinated the famed co-discoverer of natural selection throughout his life.

In an extensive analysis of Malayan swallow-tailed butterflies read before the Linnean Society in March of 1864, Wallace insisted "that not the wing of a butterfly can change in form or vary in color, except in harmony with, and as a part of the grand march of nature."[19] Wallace felt that the butterflies he encountered in the Malay Archipelago gave evidence of natural selection in action. They *also*, as he would make clearer later on, gave evidence of design and purpose in nature. Butterflies, therefore, offer an interesting glimpse into Wallace's development as an evolutionary biologist and intelligent design advocate.

That development began with the April 1869 issue of the *Quarterly Review* when he startled Charles Darwin and the rest of Darwin's entourage (especially Thomas Henry Huxley and his X Club cohorts) by invoking an "Overruling Intelligence" to account for the evolution of Homo sapiens. From that point on, and despite Darwin's deep consternation, Wallace continued to expand and develop an increasingly teleological worldview.

So when, after a lifetime of scientific investigation and inquiry, he contemplated the amazing life cycle of the butterfly, he was forced to conclude that there was some guidance behind the transformation of the modest caterpillar into the majestic beauty of its culminated form. Wallace described the process of metamorphosis thus:

Everyone knows that a caterpillar is almost as different from a butterfly or moth in all its external and most of its internal characters, as it is possible for any two animals of

---

[18] Alfred Russel Wallace, *The Malay Archipelago* (New York: Harper & Brothers, 1869), p. viii.
[19] Alfred Russel Wallace, "The Malayan Papilionidæ or Swallow-Tailed Butterflies as Illustrative of the Theory of Natural Selection," in *Contributions to the Theory of Natural Selection: A Series of Essays* (New York: Macmillan, 1871), p. 198.

the same class to be. The former has six short feet with claws and ten fleshy claspers; the latter, six legs, five-jointed, and with subdivided tarsi; the former has simple eyes, biting jaws, and no sign of wings; the latter, large compound eyes, a spiral suctorial mouth, and usually four large and beautifully colored wings. Internally the whole muscular system is quite different in the two forms, as well as the digestive organs, while the reproductive parts are fully developed in the latter only. The transformation of the larva into the perfect insect through an intervening quiescent pupa or chrysalis stage, lasting from a few days to several months or even years, is substantially the same process in all the orders of the higher insects, and it is certainly one of the most marvelous in the whole organic world. The untiring researches of modern observers, aided by the most perfect microscopes and elaborate methods of preparation and observation, have revealed to us the successive stages of the entire metamorphosis, which has thus become more intelligible as to the method or succession of stages by which the transformation has been effected, though leaving the fundamental causes of the entire process

as mysterious as before.... There is, I believe, nothing like this complete decomposition of one kind of animal structure and the regrowth out of this broken-down material which has thus undergone decomposition of the cells, but not apparently of the protoplasmic molecules to be found elsewhere in the whole course of organic evolution; and it introduced new and tremendous difficulties into any mechanical or chemical theory of growth and of hereditary transmission.[20]

Indeed, Wallace thought that birds also gave evidence of intelligent design, noting, "the bird's wing seems to me to be, of all the mere mechanical organs of any living thing, that which most clearly implies the working out of a preconceived design in a new and apparently most complex and difficult manner, yet so as to produce a marvelously successful result."[21] Wallace doubted that all of the diverse coloration of birds could be explained on the mere principle of utility, namely, that it afforded them in each and every case a survival advantage in nature. In

[20] Alfred Russel Wallace, *The World of Life: A Manifestation of Creative Power, Directive Mind and Ultimate Purpose* (London: Chapman and Hall, 1910), pp. 298, 300.
[21] Ibid., pp. 287-288.

applying this to insects like butterflies, Wallace believed that they

not only present us with a range of color and pattern and of metallic brilliancy fully equal (probably superior) to that of birds, but they possess also in a few cases and in distinct families, changeable opalescent hues, in which a pure crimson, or blue, or yellow pigment, as the incidence of light varies, changes into an intense luminous opalescence, sometimes resembling a brilliant phosphorescence more than any metallic or mineral luster …. And what renders the wealth of coloration thus produced the more remarkable is, that, unlike the feathers of birds, the special organs upon which these colors and patterns are displayed are not functionally essential to the insect's existence. They have all the appearance of an added superstructure to the wing, because in this way a greater and more brilliant display of color could be produced than even upon the exquisite plumage of birds. It is true that in some cases, these scales have been modified into scent-glands in the males of some butterflies, and perhaps in the females of some moths, but otherwise they are the vehicles of color alone; and though the diversity of tint and pattern is undoubtedly useful in a variety of ways to the insects themselves, yet it is so almost wholly in relation to higher animals and not to their own kind.…The brilliant metallic or phosphorescent colors on the wings of butterflies may serve to distract enemies from attacking a vital part, or, in the smaller species may alarm the enemy by its sudden flash with change of position. But while the colors are undoubtedly useful, the mode of producing them seems unnecessarily elaborate, and adds a fresh complication and a still greater difficulty in the way of any mechanical or chemical conception of their production.[22]

Figure 1. Alfred Wallace

[22] Ibid., pp. 302-304.

In the final analysis, Wallace concluded that if the coloration and beauty of birds couldn't be fully explained on the principle of utility, the complex process of metamorphosis and the resultant beauty and splendor of butterflies magnified the problem. If butterflies reflect "the grand march of nature," then by 1910 Wallace declared that Nature herself has told the story on their wings "like the pages of some old illuminated missal, to exhibit all her powers in the production, on a miniature scale, of the utmost possibilities of color-decoration, of color-variety, and of color-beauty; and has done this by a method which appears to us unnecessarily complex and supremely difficult, in order perhaps to lead us to recognize some guiding power, some supreme mind, directing and organizing the blind forces of nature in the production of this marvelous development of life and loveliness."[23]

This raises a question. How, if Wallace made this great departure from Darwin's methodological naturalism in 1869 by invoking an Overruling Intelligence and by 1910 was seeing clear features of design and purpose in butterflies, could he leave *unamended* an essay written several years *before* in a collection, *Contributions to the Theory of Natural Selection*, he personally compiled in 1870? If, as one recent student of Wallace has claimed, "Wallace's earlier belief in laws was replaced by his belief in ongoing providential guidance,"[24] then the answer to this question is unobtainable. However, this view misconstrues the *laws* of nature. A dichotomy needn't exist between teleology or "providential guidance" and natural law. It also misinterprets Wal-

Figure 2. Wallace's Golden Birdwing butterfly (Ornithoptera croesus). Wallace described the exhilaration of capturing this species he called "the pride of the Eastern tropics." In the forest at Batchian (one of the northern Spice Islands), Wallace discovered a shrub frequented by these stunning butterflies. "The beauty and brilliancy of this insect are indescribable," he wrote, "and none but a naturalist can understand the intense excitement I experienced when I at length captured it" (*The Malay Archipelago*, p. 300).

---

[23] Ibid., p. 323.

[24] Jakob Novák, "Alfred Russel Wallace's and August Weismann's Evolution: A Story Written on Butterflies" (PhD diss., Princeton University, 2008), p. 205.

lace, for although he never embraced Christianity, his teleological biology was a natural theology of sorts. Understood in this way, Jay Richards's observation seems germane: "if God [or an Over-ruling Intelligence] is in charge, then even where natural selection is work-ing, no variation will be literally random in the sense that most Darwinists un-derstand the word. They won't be purposeless."[25]

As we have seen, Wallace him-self said that "some guiding power" or "supreme mind" was responsible for "directing and organizing the blind forces of nature." In other words, Wal-lace did not view teleology as being at odds or in tension with natural law. The stochastic and utilitarian properties of natural selection are limiting factors that demand a design inference where those properties are lacking; natural laws and purposeful design were not either/or propositions for Wallace who certainly allowed that "the controlling action of such higher intelligences [i.e., teleological forces] is a necessary part of those laws."[26] Wallace once put it bluntly, "There are laws of nature, but they are purposeful."[27] Wallace demon-strated three propositions regarding in-telligent design in nature: 1) it needn't demand miraculous interventions; 2) law-like phenomena don't preclude teleology; and 3) intelligent design is fully consonant with science, for how else could Wallace have let his 1864 essay stand unrevised! For Wallace, as wondrous as were all species of but-terflies, the direct and special act of a creator was not required to explain the existence of every one of them, simply some guidance or teleological power to direct the common descent of Lepidop-tera. In short, evolution itself was *intelligent*.

This was not a *volte-face* for Wal-lace. Hints of teleology can be seen in his life and work early on. For example, as a young man of twenty writing late in 1843 he asked, "Can any reflecting mind have a doubt that, by improving to the utmost the nobler faculties of our nature in this world, we shall be the better fitted to enter upon and enjoy whatever new state of being that future may have in store for us?"[28] These are hardly the words of a committed ma-terialist or atheist. Then in 1856, two years before his famous Ternate letter explicating the theory of natural selec-tion, he chided his fellow naturalists for being "too apt to *imagine*, when they cannot *discover*, a use for everything in nature: they are not even content

25 Jay Richards, "Understanding Intelligent Design," in *God and Evolution,* edited by Jay Richards (Seattle: Discovery Institute Press, 2010), pp. 256-257.
26 Alfred Russel Wallace, "The Limits of Natural Selection as Ap-plied to Man," in *Contributions,* p. 360.
27 Alfred Russel Wallace, *New Thoughts on Evolution. Being the View of Dr. Alfred Russel Wallace. As Gathered in an Interview by Harold Begbie* (London: Chapman and Hall, 1910), p. 14.
28 Alfred Russel Wallace, *My Life: A Record of Events and Opin-ions* (1908; reprinted, Elibron Classic, 2005), p. 114.

to let 'beauty' be a sufficient use, but hunt after some purpose to which even *that* can be applied by the animal itself, as if one of the noblest and most refining parts of man's nature, the love of beauty for its own sake, would not be perceptible also in the works of a Supreme Creator."[29] Even then Wallace perceived limiting factors in purely naturalistic explanations and was sensing higher purpose in nature. At least one Wallace biographer, H. Lewis McKinney, has noted Wallace's call in 1863 for ecological responsibility in light of "*Creation* which we had in our power to preserve" and the hypocrisy expressed in poor stewardship notwithstanding the general belief in a direct Creator. For McKinney this is suggestive of Wallace's shifting religious beliefs well *before* his being introduced to then-fashionable spiritualism two years later."[30] The notion held by some (for example, Malcolm Jay Kottler and James Moore[31]) that Wallace's conversion to spiritualism—a belief shared by many of his scientific colleagues—explains his increasing commitment to teleology is a facile compartmentalizing and dividing of Wallace "the scientist"

from Wallace "the theist," as if the two could not coexist.

In fact it is pointless to imagine a pre-teleological Wallace and a post-teleological Wallace. There were never two Wallaces. The same Wallace who searched the laws of nature was the same Wallace who, the more he investigated them, began to perceive the harmonies behind them. Wallace's teleological world was as law-based and as unbroken as ever. Butterflies hadn't changed either, but "the grand march of nature" to which they colorfully paraded was for Wallace increasingly to the cadence of a teleological drum. In the end, butterflies taught Wallace that the seeming cacophony of atonal materialism was in fact a symphony of purpose. A new and different Wallace needn't be posited simply because he was finally able to read the "notes" written on butterfly wings.

Figure 3. This striking butterfly bearing the name *Heliconius wallacei* or

---

29 Alfred Russel Wallace, "On the Habits of the Orang-utan of Borneo," *Annals and Magazine of Natural History.* 2nd series. 17, no.103 (1856), pp. 26-32.

30 *Dictionary of Scientific Biography*, s.v., "Wallace, Alfred Russel."

31 See Malcolm Jay Kottler, "Alfred Russel Wallace, the Origin of Man, and Spiritualism," *Isis* 65, no . 2 (1974), pp. 144-192; and James Moore, *Post-Darwinian Controversies: A Study of the Protestant Struggles to Come to Terms with Darwin in Great Britain and America, 1870-1900* (Cambridge, United Kingdom: Cambridge University Press, 1979), pp. 184-190.

Wallace's Long Wing butterfly (above) is found in the Amazon River Basin. Wallace's discovery of new and different butterflies in the Malay Archipelago only continued work he had begun in the Amazon River Basin (1848-1852). "It is in the lovely butterflies that the Amazonian forests are unrivalled," he remarked, "whether we consider the endless variety of the species, their large size, or their gorgeous colors. South America is the richest part of the world in this group of insects, and the Amazon seems the richest part in South America." From his *Narrative of Travels on the Amazon and Rio Negro* (London: Reeve and Co., 1853), p. 468.

# Chapter 8

## Magic Masks of Mimicry: Vladmir Nabokov as Darwin Doubter

David Klinghoffer

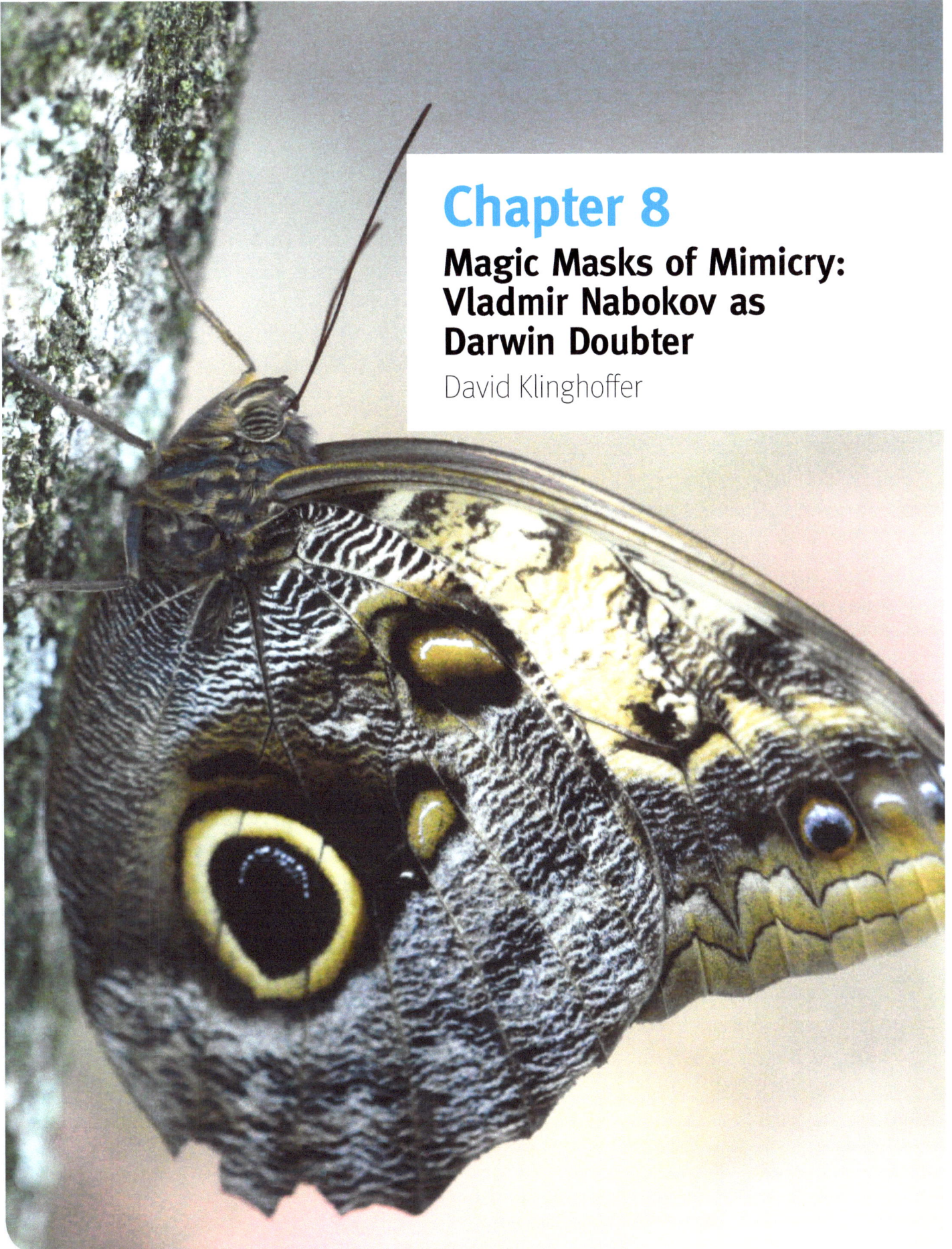

# Chapter 8
## Magic Masks of Mimicry: Vladimir Nabokov as Darwin Doubter

David Klinghoffer

Was Vladimir Nabokov (1899-1977) a fundamentalist Christian or simply a scientific ignoramus? Among Darwinists, those are the most commonly cited explanations for the strange phenomenon where people shockingly and otherwise unaccountably pipe up with doubts about Darwinian theory. The great novelist (*Lolita, Pale Fire, Pnin*) was, in his own telling, a "furious" critic of Darwinian theory. In fact, he based the judgment not on religion, to which his biographer Brian Boyd writes that he was "profoundly indifferent,"[32] and certainly not on ignorance but on decades of his scientific study of butterflies, including at Harvard's Museum of Comparative Zoology and the American Museum of Natural History.

It's interesting, then, how in media coverage of a recent story that reminded the public of his expertise in Lepidoptera, Nabokov's Darwin doubting was slyly elided. Or perhaps not slyly but cluelessly. "Nonfiction: Nabokov Theory on Butterfly Evolution Is Vindicated," read the *New York Times* headline on January 25, 2011. "Nabokov Was Right All Along," as evolutionary biologist and thumper for atheism Jerry Coyne put it in the title of a post on his blog site, Why Evolution Is True.[33] Coyne, who teaches at the University of Chicago, wrote about how he found it "really cool" that a paper Nabokov published in 1945 in the scholarly butterfly journal *Psyche* had just been confirmed by modern gene sequencing.

Nabokov's focus in the world of butterflies and moths was the *Polyommatus* blues. In the article he had speculated, based on a searching review of the blues and their genitalia, that the group invaded the Americas from Asia in a series of migrations via the Bering Strait, starting 11 million years ago. As a team writing in the *Proceedings of the Royal Society of London* now confirmed, Nabokov's speculation was prescient and, as retrospectively demonstrated, accurate in every detail.

Not mentioned, however, by newspaper or braying atheist was how Nabokov in the *very same paper* took a swipe at Darwinian theory. He wrote that "'natural selection' in its simplest sense…certainly had no direct action whatever on the molding of the genital armature….While accepting evolution as a modal formula"—describing the changing character of populations over

---

32 Brian Boyd, *Vladimir Nabokov: The American Years* (Princeton: Princeton University Press, 1991), p. 291.

33 Jerry Coyne, "Nabokov Was Right All Along," http://whyevolutionistrue.wordpress.com/2011/01/27/nabokov-was-right-all-along/.

time—"I am not satisfied with any of the hypotheses advanced in regard to the way it works."[34]

This is just another way of saying what the scientists who have signed Discovery Institute's Dissent from Darwinism list affirm ("We are skeptical of claims for the ability of random mutation and natural selection to account for the complexity of life"). Nabokov naturally accepted that life evolved in the sense of changing over vast stretches of time, but he rejected all contemporary formulations, including Darwin's, that seek to explain by what *mechanism* evolution occurs.

Writing in the same issue of *Psyche*, he goes on to say that "repetitions of structure"—the same biological structures recurring around in butterflies the world—cannot be explained as resulting from "haphazard 'convergence' since the number of coincident characters in one element, let alone the coincidence of that coincident number with a set of characters in another element, exceeds anything that might be produced by 'chance.'" This is an argument familiar today in discussions of the convergence theory advocated by Simon Conway Morris.

Nabokov's comment, while escaping the notice of Jerry Coyne and the *New*

*York Times*, was far from an isolated incidence of his public Darwin doubting. As Brian Boyd notes in his biography, *Vladimir Nabokov: The American Years*, "He could not accept that the undirected randomness of natural selection would ever explain the elaborateness of nature's designs, especially in the most complex cases of mimicry where the design appears to exceed any predator's powers of apprehension."[35]

Just how noted a lepidopterist was Nabokov? In an appreciation of his scientific work written for the magazine *Natural History*, Boyd summarized the artist's bona fides:

For most of the 1940s, he served as de facto curator of Lepidoptera at Harvard University's Museum of Comparative Zoology, and became the authority on the little-studied blue butterflies (*Polyommatini*) of North and South America. He was also a pioneer in the study of butterflies' microscopic anatomy, distinguishing otherwise almost identical blues by differences in their genital parts.

Later employed at Harvard as a research fellow in entomology while

---

[34] Reproduced in Vladimir Nabokov, *Nabokov's Butterflies: Unpublished and Uncollected Writings*, ed. Brian Boyd and Robert Michael Pyle (Boston: Beacon Press, 2000), p. 356.

[35] Boyd, *Vladimir Nabokov: The American Years*, p. 23

teaching comparative literature at Wellesley, Nabokov published scientific journal articles in *The Entomologist, The Bulletin of the Museum of Comparative Zoology, The Lepidopterists' News,* and *Psyche: A Journal of Entomology.*

As Boyd notes, Nabokov wrote "a major article," subsequently lost, "with 'furious refutations of "natural selection" and "the struggle for life."'" He completed the paper in 1941 but all that survives is a fragment in his memoir, *Speak, Memory* (1951):

The mysteries of mimicry had a special attraction for me. Its phenomena showed an artistic perfection usually associated with man-wrought things. Consider the imitation of oozing poison by bubblelike macules on a wing (complete with pseudo-refraction) or by glossy yellow knobs on a chrysalis ("Don't eat me—I have already been squashed, sampled and rejected"). Consider the tricks of an acrobatic caterpillar (of the Lobster Moth) which in infancy looks like bird's dung, but after molting develops scrabbly hymenopteroid appendages and baroque characteristics, allowing the extraordinary fellow to play two parts at once (like the actor in Oriental shows who becomes a pair of intertwisted wrestlers): that of a writhing larva and that of a big ant seemingly harrowing it. When a certain moth resembles a certain wasp in shape and color, it also walks and moves its antennae in a waspish, unmothlike manner. When a butterfly has to look like a leaf, not only are all the details of a leaf beautifully rendered but markings mimicking grub-bored holes are generously thrown in. "Natural Selection," in the Darwinian sense, could not explain the miraculous coincidence of imitative aspect and imitative behavior, nor could one appeal to the theory of "the struggle for life" when a protective device was carried to a point of mimetic subtlety, exuberance, and luxury far in excess of a predator's power of appreciation. I discovered in nature the nonutilitarian delights that I sought in art. Both were a form of magic, both were a game of intricate enchantment and deception.[36]

The novelist/lepidopterist asked, if a particular artistic subtlety is beyond the ability of a predator to perceive, how did nature select it? That sounds an awful lot like the lead-up to an argument for intelligent design.

---

[36] Vladimir Nabokov, Speak, Memory (New York: Everyman's Library, 1999), p. 88.

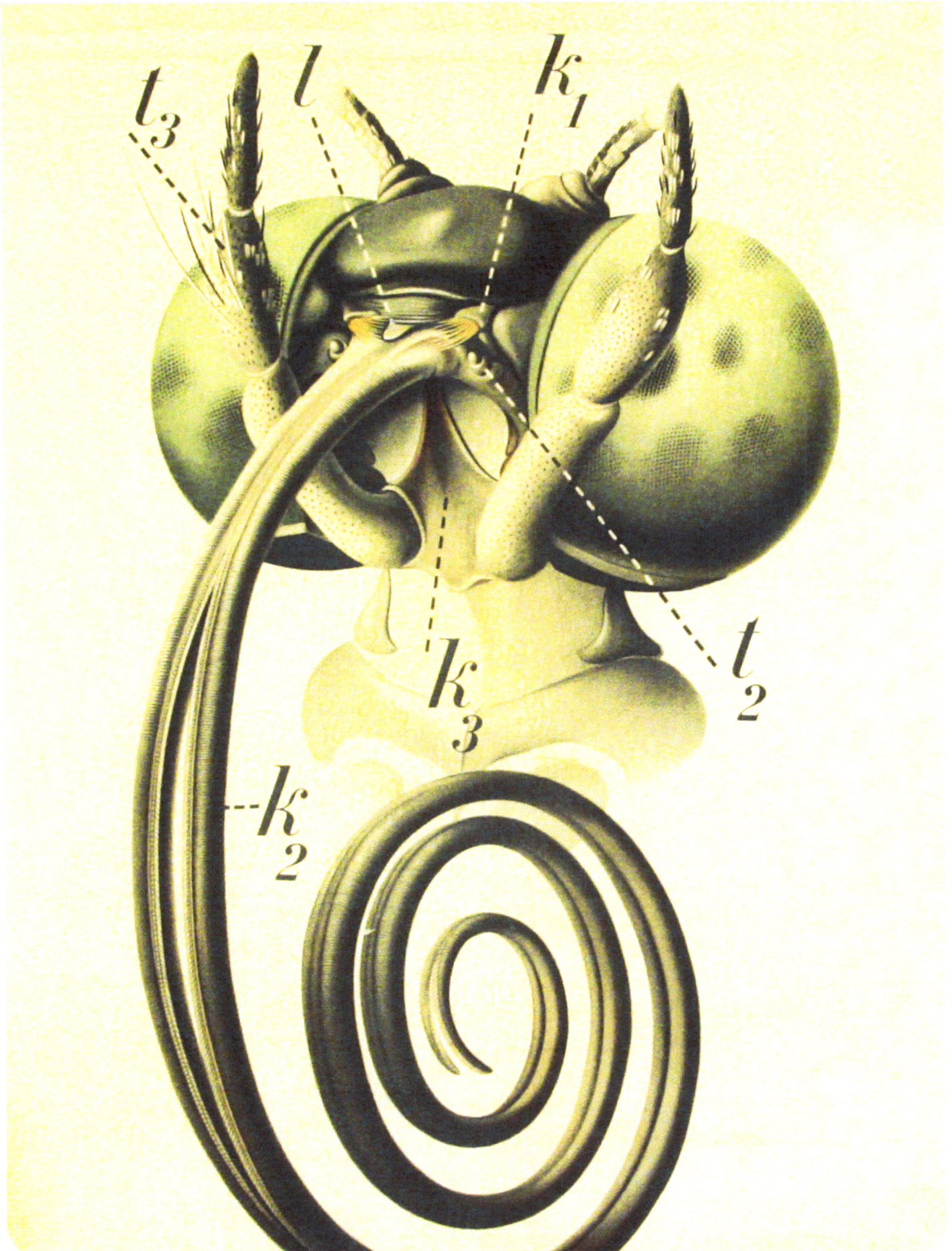

A prime source on Nabokov on design in nature is a book that Boyd edited with Robert Michael Pyle, *Nabokov's Butterflies*, collecting his scientific and literary writing on butterflies, both published and previously unpublished. Nabokov, writes Boyd, detected a "sense of design deeply embedded in nature's detail, of a playful deceptiveness behind things, of some kind of conscious cosmic hide-and-seek." Although there is no evidence that Nabokov understood the intelligence behind the design to be a deity, the situation in his mind was much as the Bible had put it: "It is the glory of God to hide a thing, but the glory of kings to search things out" (Proverbs 5:2). Mimicry among butterflies was one of those give-away clues in nature that point most clearly to hidden design.

"Only in mimicry," says Boyd, "did he suspect that the design behind things was apparent enough and explicit enough to be treated as science."[37]

In his own introductory essay to the book, lepidopterist Robert Michael Pyle writes of how Nabokov challenged his colleagues, the evolutionists Ernst Mayr and Theodosius Dobzhansky, on the nature of species. As I mentioned, he planned to issue a formal challenge, rather than an incidental one.

In 1941, he wrote in a letter to novelist and chemist Mark Aldanov that simultaneously with his work on his novel *Bend Sinister* he was "writing a work on mimicry (with a furious refutation of 'natural selection' and the 'struggle for life')."[38] What a shame that work does not survive. But in his novel *The Gift* he considered including an addendum titled "Father's Butterflies" that captures at some length the source of his doubts about Darwin.

The addendum was never published as part of the novel but a manuscript was preserved in the Library of Congress and appears, translated by his son Dimitri, in *Nabokov's Butterflies*. On mimicry, he argued that evolution never had the time to produce such wonders by a Darwinian process: "The impossibility of achieving false similarities via a gradual accumulation of corresponding traits, whether by chance or as a consequence of 'natural selection,' is proven by a simple lack of time."[39] As published (in English translation in 1963), *The Gift* recounts the narrator's memories of his father's arguments along the same lines, the "magic masks of mimicry."

He told me about the incredible artistic wit of mimetic disguise, which

---

37 Nabokov, *Nabokov's Butterflies*, pp. 19-20.

38 Ibid., p. 248.
39 Ibid., p. 223.

was not explainable by the struggle for existence (the rough haste of evolution's unskilled forces), was too refined for the mere deceiving of accidental predators, feathered, scaled and otherwise (not very fastidious, but then not too fond of butterflies), and seemed to have been invented by some waggish artist precisely for the intelligent eyes of man.[40]

In the Nabokov Archive at the New York Public Library, Boyd and Pyle discovered further unpublished notes on issues related to speciation, made in the course of preparing another paper for *Psyche* ("Notes on the Morphology of the Genus *Lycaeides*," 1944). In one note on what he called "homopsis," a kind of repetition of characteristics among species—in this case, those of the butterfly genus *Lycaeides*—he commented on the coincidence of various mimetic traits. The coincidence was of a kind that it was "impossible to explain satisfactorily either by blind accidental causes or by the blind coordination of accidents termed natural selection (even if the protective value of mimetic resemblance is proved)."[41]

Nabokov's view on the inadequacies of natural selection as a theory was no casual eccentricity, just as it was no concession to any religious faith. As a lepidopterist, he was not some dabbling dilettante but a serious scientist with a mastery of his subject and a perceptiveness that would be further demonstrated with time. This, not surprisingly, puts Nabokov scholars in a bind. They can't and don't want to minimize his scientific expertise. They can't deny the obvious reality that with regard to Darwinism, he was a flaming heretic. But neither do the prejudices of academia permit them to validate Nabokov's view as one that he could have sustained indefinitely.

In a footnote in the second volume of his biography, Boyd reassures us that—perhaps on the model of Darwin's disturbing racism—Nabokov's heresy was a product of his time and therefore excusable. Back in the day, "his position was not so unusual as it may seem now. Among professional biologists it was only in the decade 1937-1947 that what Julian Huxley called 'the evolutionary synthesis' itself evolved, and settled the differences between naturalists and geneticists that had impeded widespread acceptance, not of evolution per se, but of Darwin's own explanations of the phenomena."[42]

That's a nice try. Stephen Jay Gould tried a similar tack in an essay ("No Science Without Fancy, No Art With-

[40] Ibid., p. 178.
[41] Ibid., p. 310.

[42] Boyd, *Vladimir Nabokov: The American Years*, p. 23.

out Facts: The Lepidoptery of Vladimir Nabokov"), arguing that "When Nabokov wrote his technical papers in the 1940s, the modern Darwinian orthodoxy had not yet congealed, and a Nabokovian style of doubt remained quite common among evolutionary biologists."[43]

Yet while leading evolutionists like Mayr and Dobzhansky were synthesizing the modern synthesis, Nabokov wasn't lost on a desert island or isolated on an Alpine peak. He was present at the idea's birth, at Harvard and the American Museum of Natural History, and was a contemporary of the very leaders in the scientific movement that gave Darwinism its "neo" form. In their book *Nabokov's Blues: The Scientific Odyssey of a Literary Genius*, lepidopterist Kurt Johnson and journalist Steve Coates note that Nabokov followed Mayr and Dobzhansky's work in particular "with critical interest."[44]

Yet Johnson and Coates too sympathize with the Gould defense, explaining that nowadays everyone in the field would know "how actions of genes in specific populations drive evolution toward sometimes fantastic but still mechanistic external resemblances." But even if Nabokov had been elsewhere than at Harvard at precisely the

time such ideas were in the process of winning orthodox approval, it's hard to see how the neo-Darwinian revelation would dissolve the dilemmas that he perceived.

Johnson and Coates also consider the possibility that while Nabokov was not influenced by religious belief, perhaps other philosophical views shaped his doubts about Darwin. A lepidopterist at Yale and friend of Nabokov's, Charles Lee Remington, suggested a parallel between Nabokov on mimicry and the thought of Russian esoteric philosopher P.D. Uspensky (1878-1947). A vitalist, write Johnson and Coates, Uspensky "nourished the Aristotelian belief that there is a driving intelligence behind nature and that natural processes, like mimicry, are developed by this intelligence toward a desired end.... While Uspensky's often magical view of the world suggests some of the layered realities Nabokov himself created in works like *Pale Fire*, it seems unlikely that Nabokov the scientist would have put much faith in Uspensky's overall worldview."[45]

It's not clear whether Nabokov ever read Uspensky's work. But as a young man he is known to have read another vitalist and anti-materialist, Henri Bergson, with attention and admiration. Brian Boyd in the first volume of his

---

[43] Quoted in Kurt Johnson and Stephen L. Coates, *Nabokov's Blues: The Scientific Odyssey of a Literary Genius* (Hanover, N.H.: Zoland Books, 2000), p. 328.
[44] Ibid., p. 50.

[45] Ibid., pp. 328-329.

biography, *Vladimir Nabokov: The Russian Years*, notes that Nabokov went even further than Bergson in his anti-Darwinism—which surfaced, at least implicitly, as early as his 1924 short story "The Dragon."

It's possible that Nabokov was influenced decisively and led astray by other people's ideas, or possibly by symbolic, metaphysical meanings he attached to butterflies. They turn up frequently as a motif in his fiction, often denoting the soul's liberation at death—a metamorphosis like the one that takes the caterpillar through the chrysalis to its transmogrification in the form of a butterfly. The name of the peer-reviewed scientific journal where he published many of papers, *Psyche*, means in Greek "soul," "life," "ghost," "self," and "butterfly."

"We are the caterpillars of angels," he wrote in a 1923 poem.[46] Charles Lee Remington believed his old friend was so committed to his metaphysics that even in the face of the fully articulated neo-Darwinian doctrine, had he remained professionally active as a scientist, Nabokov might well have refused to relent.

But all such speculations are really intended to avoid confronting the possibility that his view was just what it seemed to be—honestly attained with eyes open, by sincerely, intelligently, and openly considering the relevant evidence of life, science side by side with art. Darwinists are invested, always, in explaining away and deconstructing other people's evolutionary doubts. Nabokov's science has already been vindicated on one relatively arcane question. Why not, too, on what is—for scientists and non-scientists—the biggest question there could be?

---

46 Nabokov, *Nabokov's Butterflies*, p. 109.

# Chapter 9
## Darwin vs. Beauty:
## Explaining Away the Butterfly

Jonathan Witt

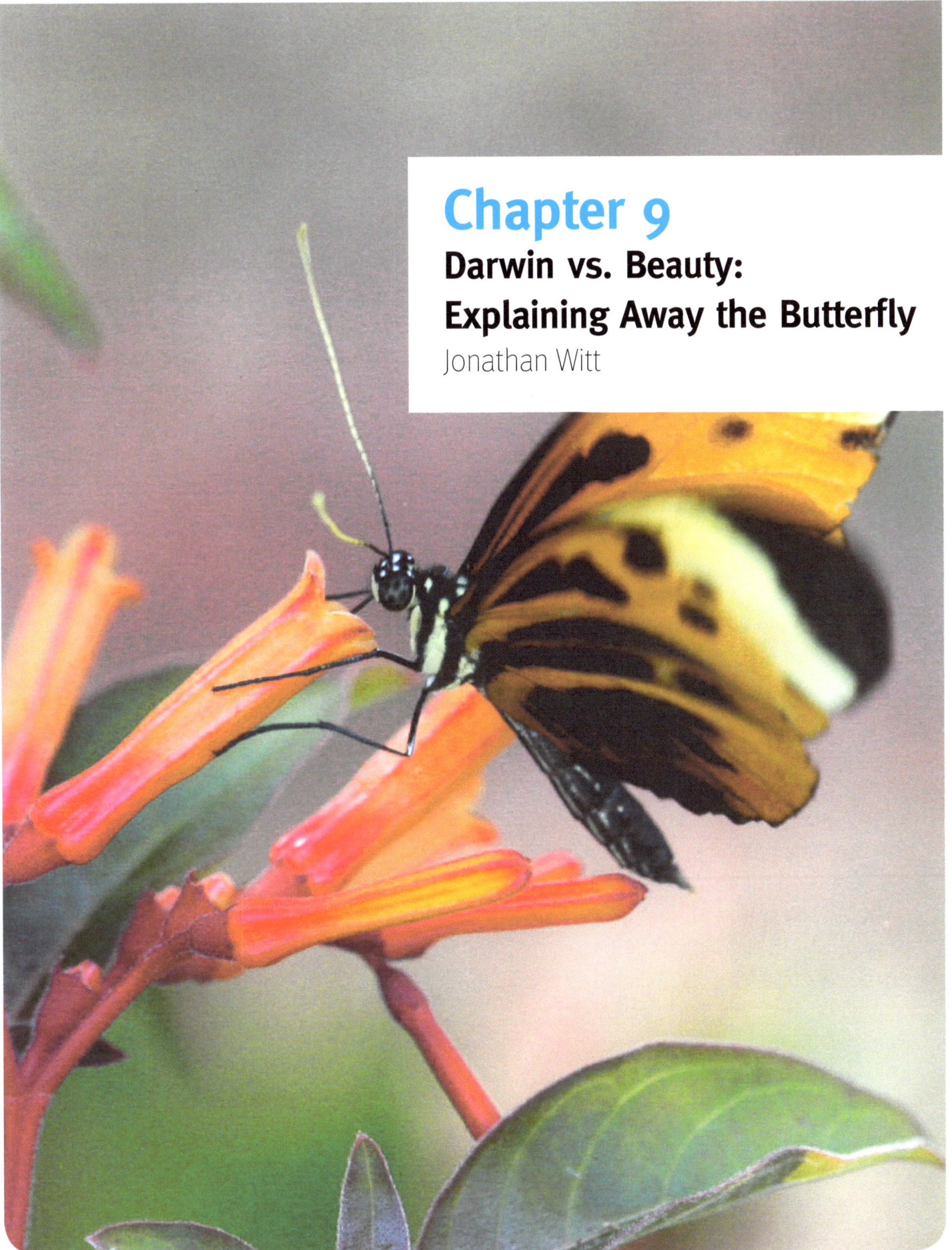

# Chapter 9
# Darwin vs. Beauty: Explaining Away the Butterfly

Jonathan Witt

There is nothing necessarily illogical about seeking to "explain away" something, since the something in question may be an illusion; but a first step in understanding Darwinism's response to the beauty of the butterfly—and to beauty generally—is to recognize that Darwinism does seek to explain away our experience of it. In particular, it seeks to explain away our sense that beauty in some way connects us to the transcendent.

To his credit, Charles Darwin recognized there were instances of extravagant natural beauty that outstripped the explanatory power of Darwinian natural selection, so in *The Descent of Man* he developed his theory of sexual selection to fill the explanatory gap. There Darwin argued, in essence, that the peacock has an extravagant tail, Shakespeare an extravagant gift for spinning tales, and Mozart an extravagant ability to compose, the better to attract a mate.[47]

His explanation, while scientific in its orientation, was part of a larger philosophical project known as reduction—the idea that the best way to understand something is to identify its material parts, and to do so at lower and lower levels (e.g., from traits to cells to molecules to atoms, and so on). At its most extreme, reductionism views things as ultimately just the sum of their parts. Thus, to a Darwinian reductionist, the grace and beauty of the butterfly or the songbird or the poet ultimately spring from some advantage this beauty lent the creature and its ancestors for survival and reproduction ("survival of the fittest").

And at this point Darwinism is only getting warmed up. Darwinian reductionism is the great equalizer, boiling all of life down to either natural selection or sexual selection, and beneath that, to genetics. "Now they swarm in huge colonies, safe inside gigantic lumbering robots, sealed off from the outside world, communicating with it by tortuous indirect routes, manipulating it by remote control," explains evolutionary apologist Richard Dawkins. "They are in you and in me; they created us, body and mind; and their preservation is the ultimate rationale for our existence. They have come a long way, those replicators. Now they go by the name of genes, and we are their survival machines."[48]

If we imagine that the higher things in life are somehow exempt from this

---

47 Charles Darwin, *The Descent of Man, and Selection in Relation to Sex* (Princeton: Princeton University Press), 1981.

48 Richard Dawkins, *The Selfish Gene* (New York: Oxford University Press, 1976), pp. 19-20.

reductionist acid, Harvard socio-biologist Edward O. Wilson sets us straight. He leads into the matter by suggesting that when humans have grown wiser, the human mind "will be more precisely explained as an epiphenomenon of the neuronal machinery of the brain. That machinery is in turn the product of genetic evolution by natural selection acting on human populations for hundreds of thousands of years in their ancient environments."[49] A bit later he adds, "The social scientists and humanistic scholars, not omitting theologians, will eventually have to concede that scientific naturalism is destined to alter the foundation of their systematic inquiry by redefining the mental process itself."

As for the beauty of artistic genius, the "sensuous hues and dark tones have been produced by the genetic evolution of our nervous and sensory tissues," he writes. "To treat them as other than objects of biological inquiry is simply to aim too low."[50]

Wilson puts a brave and noble face on his recommended approach, implying as he does that his opponents are aiming "too low." But what exactly is high about Darwinian reductionism? Treating artistic beauty as a mere by-product of evolution doesn't lead to a higher, deeper or nobler understanding of art. It undermines the very foundation for saying anything is noble or low or wicked.

Think about some of the great poems, paintings or novels. They probe the world of flesh and blood, but at the same time they draw us into things spiritual: the sublime and the ridiculous; love, heroism, and envy; good and evil. But if Darwinism is right, some of our ancestors had an evolutionary mutation that caused them to imagine that a spiritual dimension—including things like nobility—actually exist. Since the illusion made them better at surviving and reproducing, the mutation passed from one generation to the next in a growing population of deluded ancestors, creatures who worked out their delusion in everything from poetry to painting to music. So goes the story of Darwinian reductionism.

One might respond, "Well, that's just the price of honest, unflinching rational investigation." But the conclusion is far from rational, and for at least a couple of reasons. First, the theory of sexual selection moves rather than solves the problem of extravagant beauty in the biological realm. Consider the tail of the peacock. Their enormous tails slow them down, making it easier for predators to catch them. Darwin's complementary theory of sexual selection says that peahens are attracted to large tail feathers (or

49 Edward O. Wilson, *On Human Nature* (Cambridge: Harvard University Press, 1978), pp. 195, 204.
50 Wilson, *On Human Nature*, p. 11.

more specifically, to the abundance of bright blue spots on the tail feathers), and they use these as a selection criterion when choosing a mate. The problem is this: Now there's another trait to be explained besides the enormous tail feathers of the peacock: namely, the tendency of peahens to choose peacocks with impractically large tail feathers. According to the Darwinian story of natural history, this trait wasn't created by an intelligent designer; it emerged gradually by natural selection. But why would nature tend to select peafowl that prefer large tail feathers on their peacocks?

Imagine you have a population of peafowl. Some of the peahens select their mates in the ordinary way—according to how fast the peacocks can take off, by how well they can handle themselves in a fight with other peacocks, that sort of thing. But a curious cluster of genetic mutations bestowed on one peahen the gift of appreciating artistic effects, including an impractical thirst for big, beautiful plumage. Consequently, she and some of her female offspring start selecting peacocks with bigger tail feathers. The question is: why would natural selection prefer these pea hens with their impractical disposition over pea hens with survival-oriented selection criteria, criteria that would help their offspring better survive amidst a host of predators searching for dinner? Darwin's theory of sexual selection doesn't answer this question. Thus, it moves rather than solves the problem of the impractical peacock tail.

A second and more wide-ranging way that Darwinian reductionism is less than fully rational is in its commitment to the principle of methodological materialism. This is the investigative rule which says that investigators may consider only theories fully consistent with atheism. (It's not usually described this starkly, but that's what it boils down to.) According to the dictates of methodological materialism, if the extravagant beauty of butterflies or birds, if the origin of life or the universe or the fine tuning of the laws and constants of nature, if any of these features of our world points strongly toward a creative intelligence beyond the purely material, the flow of the evidence must be resisted.

This is what passes for scientific rationality in our age. But it isn't hard-nosed realism at all; it's priggish dogmatism. It's the man in the seat beside you at a Beethoven concert insisting that everything you're hearing is only so many notes, which are only so many sound waves, which are only so many perturbations among so many gaseous molecules amidst the machinery of your ear drum, the whole experience a curious stew of physics and sexual selection working its soulless magic upon a delusional audience. The prig has talked all about the parts but has missed the whole, has missed the genius.

In the midst of such reductionism, the *Dictionary of the History of Ideas* strikes a hopeful note: "Although we are reminded that the man of the second half of the twentieth century no longer believes in geniuses, they can hardly be abolished by an act of 'cultural will.'"[51]

The evidence of artistic genius, whether human or natural, remains all around us. The evidence that we live in a world not only red in tooth and claw, but also overflowing with beauty and meaning—this too remains all around us.

Perhaps, then, we should take the existence of beauty in nature as a starting point, since we're much more directly acquainted with *that* than with the truth of the various theoretical attempts to explain them away. We need only leave the flatland of Darwinian reductionism to see them for what they are. When we do, we will find a richer explanation not only for the beauty of the butterfly, but also for the origin of species. In doing so we will have left the world of Aldous Huxley's ironically titled *Brave New World*—with its utilitarian pleasure seekers oblivious of the transcendent—and will have returned to the far richer universe of meaning and wonder that led William Shakespeare's Miranda to exclaim to her father Prospero, "How many goodly creatures are there here! … O brave new world!"

Does sex come into it? Of course. But that too is a work of genius.

---

51 "Genius: Individualism in Art and Artists," *The Dictionary of the History of Ideas: Studies of Selected Pivotal Ideas*, Vol. 2, p. 311, ed. Philip P. Wiener, http://xtf.lib.virginia.edu/xtf/view?docId=DicHist/uvaGenText/tei/DicHist2.xml;chunk.id=dv2-36;toc.depth=1;toc.id=dv2-36;brand=default (originally New York: Charles Scribner's Sons, 1973-74). Lowinsky quoted from E.E. Lowinsky, "Musical Genius—Evolution and Origins of a Concept," *The Musical Quarterly* 50 (1964), pp. 321-40.

# Contributors

**Lad Allen** is co-founder of Illustra Media, a motion picture production company based in Southern California. Lad has produced, written, and directed more than 100 films and received numerous international awards. Illustra's documentary trilogy—*Unlocking the Mystery of Life*, *The Privileged Planet*, and *Darwin's Dilemma*—has helped popularize the scientific case for intelligent design in the universe. These films have been translated into more than two dozen languages and distributed throughout the world. Illustra's most recent documentary is *Metamorphosis: The Beauty and Design of Butterflies*, which was released during the summer of 2011.

**Bernard d'Abrera** is an Australian entomological taxonomist and philosopher of science. His *magnum opus* comprises a series of works forming a synoptic reference to the true butterflies, hawk moths and saturniid moths of the world, which is based largely on the collections of the Natural History Museum in London, and other worldwide museums, public and private. The works comprise taxonomic text of over 4 million words, illustrated with over 66,000 colored figures, over ap-proximately 7,500 pages. (See the **Hill House Publishers** website, www.hill-house-publishers.com for a list of titles, regions and families treated.) D'Abrera has described several new genera as well as over 100 new species and sub-species. The D'Abrera's Tiger, *Parantica dabrerai*, an Indonesian butterfly species, is named for him, as is *Gnathothlibus dabrera*, a species of Indonesian moth.

**Michael A. Flannery** is a Discovery Institute Fellow and the author of *Alfred Russel Wallace: A Rediscovered Life* (2011) and the editor of *Alfred Russel Wallace's Theory of Intelligent Evolution: How Wallace's World of Life Challenged Darwinism* (2008). Flannery developed the content for the educational website **www.alfredwallace.org**, and he appears in the new documentary *Darwin's Heretic*. Flannery is Associate Director for Historical Collections at the Lister Hill Library of the Health Sciences and a Professor at University of Alabama at Birmingham (UAB). He has published extensively in medical history and bioethics, winning the Edward Kremers Award in 2001 for distinguished writing by an American from the American Institute of the History of Pharmacy and the 2006 Publications Award of the Archivists and Librarians in the History of the Health Sciences.

**Ann Gauger** is a Senior Research Scientist at the **Biologic Institute**. (www.biologicinstitute.org) She received a BS in biology from MIT, where she studied bacterial genetics. Her PhD at the University of Washington Zoology department expanded her horizons to include invertebrate zoology, animal physiology, cell biology and development; it was there that she first encountered the riddle of metamorphosis, while working with *Drosophila* imaginal disks. As a post-doctoral fellow at Harvard she cloned and characterized the *Drosophila* kinesin light chain gene. Now she has come full circle to focus on the origin, organization and operation of metabolic pathways in bacteria. Her research has been published in *Nature, Development*, the *Journal of Biological Chemistry*, and *Bio-Complexity.*

**David Klinghoffer** is a Senior Fellow at the Discovery Institute. His most recent book, a collaboration with Senator Joe Lieberman, is *The Gift of Rest: Rediscovering the Sabbath Day* (Howard Press/Simon & Schuster, 2011). His other books include *Why the Jews Rejected Jesus: The Turning Point in Western History* (Doubleday, 2005), *The Discovery of God: Abraham and the Birth of Monotheism* (Doubleday, 2003), and *The Lord Will Gather Me In* (Free Press/Simon & Schuster, 1999). Klinghoffer has been a literary editor and senior editor of National Review. His articles and reviews have appeared in the *New York Times, Los Angeles Times, Wall Street Journal, Washington Post, Weekly Standard,* and elsewhere.

**Paul Nelson** is a philosopher of biology who received his PhD in evolutionary theory and the philosophy of science from the University of Chicago (1998). He is currently an adjunct professor in the MA Program in Science & Religion at Biola University, and a Fellow of the Discovery Institute. His research interests include the relationship between developmental biology and the history of life, the theory of intelligent design (ID), and the interaction of science and theology. He has published chapters in many anthologies dealing with evolution and ID, and speaks frequently on these topics at colleges and universities in the United States and abroad.

**Jonathan Witt**, PhD, is a Senior Fellow with Discovery Institute's Center for Science and Culture and co-author of *A Meaningful World: How the Arts and Sciences Reveal the Genius of Nature* (2006) and *Intelligent Design Uncensored* (2010). He has written or co-written scripts for three documentaries that have appeared on PBS, including *The Privileged Planet: The Search for Purpose in the Universe*. He has written on aesthetics for *Literature and Theology* and *The Princeton Theological Review*. His essays on Darwinism, intelligent design and worldview have appeared in such places as the *Seattle Times,* the *Kansas City Star, Philosophia Christi, Crisis* and *Touchstone*. His narrative writing has appeared in the journals *Windhover* and *New Texas*.

## Photo Credits

**Unless otherwise noted, photos in this book are © Illustra Media. All Rights Reserved.**

**Chapter 1:** Figure 1, Fotolia/© Tom Hirtreiter.

**Chapter 4:** Figures 1 and 2, Paul Nelson; Figure 3, Ann Gauger, adapted and redrawn from information provided by James Truman and Lynn Riddiford, "The origins of insect metamorphosis," *Nature* 401 (1999), p. 447.

**Chapter 5:** Butterfly photos, chapter title page and p. 46, © Bernard d'Abrera and Hill House Publishers.

**Chapter 6:** Photo, p. 54, © Bernard d'Abrera and Hill House Publishers.

**Chapter 7:** Figure 1, Public Domain, from Alfred Russel Wallace, *My Life: A Record of Events and Opinions*, vol. 1. Figures 2 and 3, Wikimedia Commons (reprinted under the Creative Commons Attribution-Share Alike 3.0 Unported license).

Published by Discovery Institute Press,
208 Columbia Street, Seattle, WA 98104,
United States of America.
www.discoveryinstitutepress.com
©2011 by Discovery Institute. All Rights Reserved.
www.discovery.org/csc

# Where does your information come from?

Find out more about intelligent design by
visiting these free online resources.

CENTER FOR
SCIENCE
& CULTURE

*A Program of Discovery Institute*

## www.IntelligentDesign.Org

A portal to more information about
intelligent design, including free
email newsletters.

## www.EvolutionNews.Org

Daily news and analysis about the
debate over Darwin and
intelligent design.

## www.IDtheFuture.Com

Podcasts about evolution and
intelligent design, including interviews
with scientists.

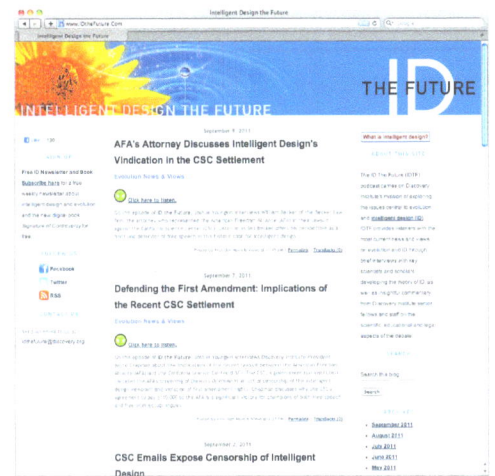

208 Columbia Street  |  Seattle, WA 98104  |  206.292.0401

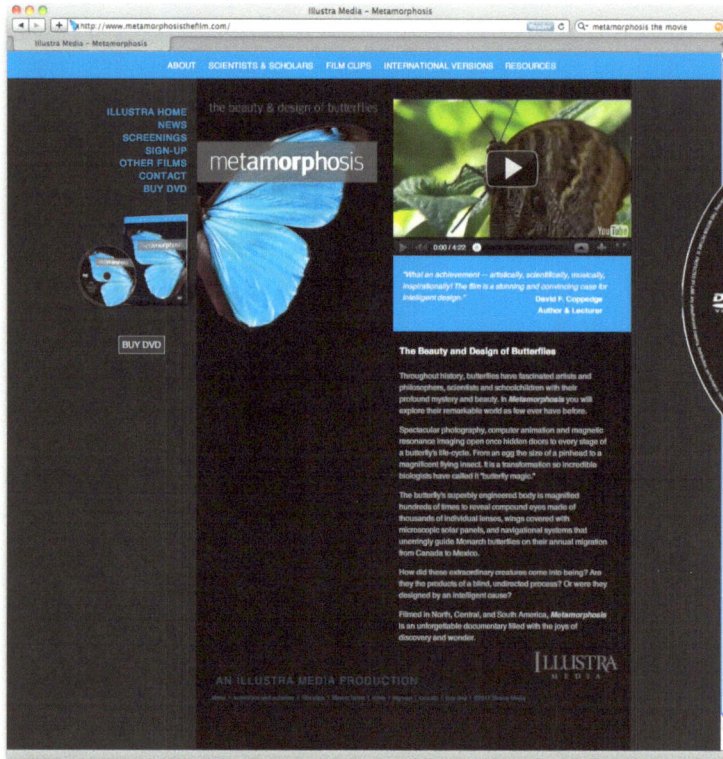

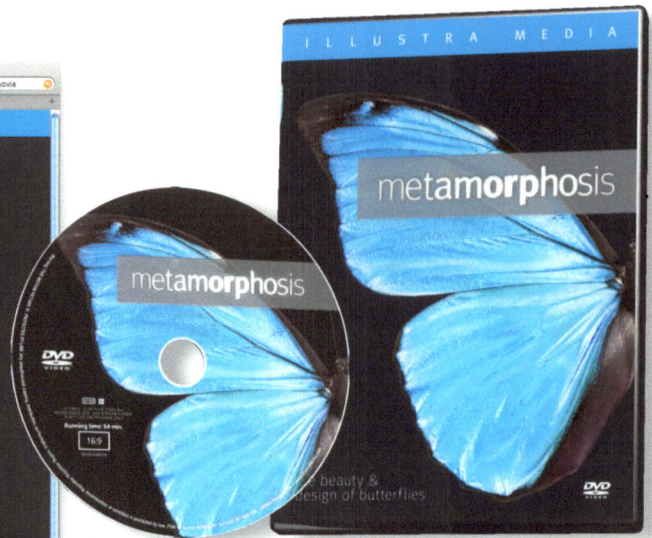

# Hill House Publishers, Ltd.

http://www.hillhouse-publishers.com/

Hill House Publishers™
Melbourne & London

*Butterflies of the World* (1961-2010) by Bernard d'Abrera is the sole modern multi-volume work by a single author/illustrator referencing the true butterflies (Papilionoidea) of the world. It is a comprehensive yet synoptic record of the entire butterfly fauna, listing every known species and many new species and races, since the famous work of Adalbert Seitz of nearly 100 years ago.

London Office Tel: +44 (0)7531 646 144 | Email: enquiries@hillhouse-publishers.com

Melbourne Office Tel: +613 9751 1141 | Email: bfly@clara.co.uk

9 781936 599035